The Logica Yearbook 2009

The Logica Yearbook 2009

edited by

Michal Peliš

© Individual author and College Publications 2010. All rights reserved.

ISBN 978-1-84890-009-7

College Publications
Scientific Director: Dov Gabbay
Managing Director: Jane Spurr
Department of Computer Science
King's College London, Strand, London WC2R 2LS, UK

http://www.collegepublications.co.uk

Original cover design by Laraine Welch
Printed by Lightning Source, Milton Keynes, UK

All rights reserved. No part of this publication may be reproduced, stored in a retrieval system or transmitted in any form, or by any means, electronic, mechanical, photocopying, recording or otherwise without prior permission, in writing, from the publisher.

Preface

The tradition of the *Logica* annual symposia goes back more than twenty years. From the beginning, the conferences have been organized by the Institute of Philosophy of the Czech (formerly Czechoslovak) Academy of Sciences. Since 1997 the symposia have been followed by a publication in *The Logica Yearbook* series. One central principle of the 'publication policy' of this series is that the volume has to appear before the beginning of next year's conference.

You now hold in your hands the thirteenth volume of the series. *The Logica Yearbook 2009* brings together various texts from mathematical and philosophical logic, the history and philosophy of logic, and natural language analysis. The aim of the conference, and of this book series, is to cover a broad field of logical topics and to promote dialogue among experts in the various branches of logic. The published papers were selected from contributions presented and discussed at the conference *Logica 2009*, which took place from June 22 to 26 in the former Franciscan monastery of Hejnice, the Czech Republic. As has become usual for the *Yearbook* series, the articles have not been sorted by subject, but are ordered alphabetically by author and it is up to the reader to pick and choose.

Both the *Logica* symposium and this book are the result of the joint effort of many people, who deserve our warmest thanks. Among them is Pavel Baran, the director of the Institute of Philosophy who has kindly supported the work of the main organizers – Timothy Childers and Vladimír Svoboda. The conference was promoted also by the Grant Agency of the Czech Republic, which provided significant support by financing grant project no. 401/07/0904. The organization of the conference would have been impossible without the work of Petra Ivaničová both prior to and during the proceedings. Our pleasant stay in Hejnice Monastery, where we spent five days not only in the lecturing hall but also in many informal discussions during the breaks, lunches and social events, was made possible due to the efforts of Father Miloš Raban and the staff of the monastery. Special thanks also go to the Bernard Family Brewery of Humpolec, traditional sponsor of the social program of the symposium. I would also like to thank to Marie Benediktová Větrovcová for the layout of this volume. Many thanks go to College Publications and its managing director Jane Spurr. Last but not least I would like to thank all the conference participants and authors of the articles for their exemplary cooperation during the editorial process.

Prague, May 2010

Michal Peliš

Contents

Doroteya Angelova
Logical and Epistemological Significance of Inconsistent and Incomplete Worlds and Their Applications   5

Jc Beall
On truth, abnormal worlds, and necessity   17

Roberto Ciuni
From Achievement Stit to Metric Possible Choices   33

Catarina Dutilh Novaes
Surprises in Logic   47

Marie Duží
Tenses and truth-conditions: a plea for if–then–else   63

Christian G. Fermüller
Some Critical Remarks on Incompatibility Semantics   81

Bjørn Jespersen
From $(AB)a$ infer A^*a   97

Annika Kanckos
Gentzen's Consistency Proofs for Arithmetic   109

John T. Kearns
What is natural about natural deduction   121

Katarzyna Kijania-Placek
Descriptive Indexicals and the Referential/Attributive Distinction   133

Hans Lycke
On Relevance Conditions for Asserting Disjunctions   143

Michal Peliš and Ondrej Majer
Logic of Questions from the Viewpoint of Dynamic Epistemic Logic   157

Andreas Pietz
    Tales of Explosions   173

Tomasz Placek
    On Attempting   183

Martin Pleitz
    "This sentence" is indexical. The indexical variant of the Liar paradox and McTaggart's paradox   195

Stephen Read
    The Validity Paradox   209

Greg Restall
    Always more   223

Christoph Roschger
    Evaluation Games for Shapiro's Logic of Vagueness in Context   231

Igor Sedlár and Juraj Podroužek
    A New Notion of Meaning Connection and the Logic of Simple Processes   247

Sebastian Sequoiah-Grayson
    Ajdukiewicz Functions and Basic Inference   259

Hartley B. Slater
    Ontological Discriminations   273

Shawn Standefer
    Philosophical Aspects of Display Logic   283

Vítězslav Švejdar
    Decision Problems of some Intermediate Logics and Their Fragments   297

Logical and Epistemological Significance of Inconsistent and Incomplete Worlds and Their Applications

Doroteya Angelova*

> *Only those who attempt the absurd ...*
> *will achieve the impossible ...*
> M. C. Escher

The aim of this paper is to show the importance of inconsistent and incomplete worlds (often called impossible worlds) in the field of logic and epistemology as well as their different applications which could also permit to be expressed more explicitly their characteristics.

These worlds have different interpretations which depend on the usage and on the underlying philosophical positions. In the different conceptions *the ways can't be* are some of the following constructs: worlds (Restall, Zalta, Varzi, Priest), situations (Barwise, Perry, Zalta, Restall), state of affairs (Vander Laan, Mares), states (Barwise, Restall), indices (Mares) and etc.[1]

1 Inconsistent and incomplete worlds in respect to logic and logical semantics

Inconsistent and incomplete worlds are generally used to be eliminated on a semantic level the "paradoxes" of implication from classical logic (namely these formulas which do not preserve the content relation between the premises and the conclusion) and respectively to be avoided the triviality of logical consequence. The last can be presented by the formulas below:

$$1.\ A \wedge {\sim}A \vDash B \quad \text{and} \quad 2.\ B \vDash A \vee {\sim}A.$$

It is known that to deny a formula for consequence, we need a "case" where the antecedent is true and the consequent is false. Thus

---

*Thanks to the organizers of LOGICA2009 for the grant they awarded me for participation in the conference.

[1] See (Restall, 1997), (Beall & Restall, 2006), (Zalta, 1997), (Varzi, 1997), (Priest, 1997), (Barwise, 1997), (Perry, 1998), (Vander Laan, 1997), (Mares, 2004).

to deny the above "paradoxical" formulas - we need a "case" or a "world" where $A \wedge \sim A$ is true (namely, an inconsistent world) and $A \vee \sim A$ is false (namely, an incomplete world). In the first of these worlds the law of non-contradiction is not valid and two contradictory sentences hold while in the second one - the law of excluded middle is invalid and a sentence and its negation are simultaneously false. So, in the most interpretations impossible worlds are these ones where the laws of non-contradiction or of excluded middle are not valid.

There are different semantics of impossible worlds developed to achieve the above conditions. The most significant logics that operate with these worlds are paraconsistent and relevant logics both of which rely on paraconsistent notion of inference. But contrary to the first one, relevant logic uses not only inconsistent but also incomplete worlds which determine my favorable attitude towards it.

In this regard I agree with Beall and Restall that since the essence of logic is the preservation of truth in all cases, which is the nature of logical consequence, the different logics appear as different explications of these cases (Beall & Restall, 2006, p. 35). In accordance with their thesis that relevant logic is well motivated, giving an own sense of the notion of "case" appropriate for a substitution in the definition of logical validity (Beall & Restall, 2000, p. 476–477), I will try to give some examples for "cases" which impossible worlds present in the definition of validity and how they may be useful for epistemological work.

For this aim let's firstly mention a few words for negation in relevant logic.

The acceptance of inconsistent and incomplete worlds imposes different conditions for negation. The relational semantics of relevant logic includes in one of its variants a unary relation $*$ on worlds (Routley star) such that $\alpha = \alpha^{**}$ and $\neg A$ holds at α iff A does not hold at α^*. It turns out that for every world α there is a correlative one α^* where the negation of the proposition is true in α iff the proposition is false in the correlative world α^*.

The following table (see the picture on page 7) will visualize the formal relationship between the worlds α and α^*, regarding α^* as a correlative world of α (Angelova, 2004, p. 202). It can't be used to compute the truth values of the statements.

Every row from the two white columns (in the middle) forms any α world and respectively every row from the hatched columns (the

Inconsistent and Incomplete Worlds

$\sim A$	A	$\sim A$	A
0	1	1	0
0	1	0	1
1	0	1	0
1	0	0	1

first and the fourth ones) is the correlative α^* world. The values of propositions A and $\sim A$ are written down, regarding them as two different variables in order to be investigated all of their properties. Thus the first row in white is an inconsistent world α (because A and $\sim A$ are simultaneously true), the second and the third rows are consistent α worlds, whereas the forth one is incomplete. As can be seen every internal world α has a correlative external one α^* obtained through the classical interpretation.

How these truth values can be explained according to the syntax of relevant logic? [2] When A and $\sim A$ are simultaneously at a

---

[2] Actually it seems to me reasonable to use the classical tableau method to identify relevant tautologies which contain only one sign of implication. For the aim when there is A and $\sim A$ at a formula they could be regarded (and have values) as different variables (see the picture on page 7); other requirement will concern the implication — the last must not be expressed with negation and disjunction and respectively with negation and conjunction as in fact is the case in relevant logic, because otherwise modus ponens with relevant implication will be able to be transformed into disjunctive syllogism. Then the disjunctive syllogism $\sim p \wedge (p \vee q) \to q$ will be regarded (using the classical tableau method) as $z \wedge (p \vee q) \to q$ which of course is not a classical tautology (and will not be a relevant one, too); the same holds also for the other "paradoxical" formulas. But further investigations on this matter could change my affinity to this method. As a matter of fact these conditions correspond to Anderson and Belnap' idea from E_{fde} where in the assessment of relevant tautologies using their disjunctive and conjunctive forms the presence of A in these forms is indifferent to the presence of $\sim A$ in

formula which is a non-relevant one are used inconsistent and incomplete worlds in order to be denied such kind of formulas. Contrary, if only one of these variables is at a formula or the both are simultaneously at a formula, which is a substituted case of a relevant one, then they have to be regarded according to the classical interpretation of negation and respectively there will not be any impossible worlds.

Thus, in classical conditions every world coincides with its correlative world (see the second and third rows) whilst it is not the case in the worlds where the laws of non-contradiction and of excluded middle are not valid (the first and the fourth rows).

The expressed conditions of negation find interesting interpretations in situation semantics which became workable tool for logic and is used successfully in the field of impossible worlds. Relevant logic often interprets inconsistent and incomplete worlds as situations. The last sometimes play the role of informational theoretical places and channels (Mares & Meyer, 2001, p. 289). The specific three-placed accessibility relation between worlds (interpreted as fusion, or channels, or links and so on) roughly speaking, represents any content connection between the propositions and therefore prevent from trivial inferences. The situations, as Perry notes (Perry, 1998, p. 669–670), contrast to the worlds in respect to the truth values - while the last determine the truth value of every proposition (and have an answer to every question), the situations correspond to limited parts of reality.

Then what happen when the situations are incomplete? They can be interpreted as restricted parts of world because of which they may leave some utterances undetermined (Beall & Restall, 2006, p. 49–50). On the other hand it is often met invaluable statements like "The present king of France is bald-headed". Thus the usage of incomplete situations may give the opportunity to be avoided a valuation of propositions which do not concern us.

There are also many epistemological reasons to accept the duality of incomplete worlds, namely, the inconsistent ones. Their significance usually appears when the subject's information is insufficient or when comes from different unreliable sources. The same holds also when there are different incompatible utterances (or beliefs) which, if are

---

the same forms (Anderson & Belnap, 1975, p. 154–158). The last sometimes is a warrant A and $\sim A$ at a formula to be regarded as different variables in order to be investigated if this formula is a substituted case of a real relevant tautology or is not from this kind and then is respectively a "paradoxical" formula.

regarded independently or quarantined in any consistent pairs, may seem possible, but when are conjoined only in one state (or story) are impossible. Something similar happens when, as Restall notes, are imposed two maps that inconsistently describe the landscape, or are connected two stories, which inconsistently describe a situation (Restall, 1997, p. 586). In order to theorize on such epistemic situations, an inconsistent theory with paraconsistent notion of inference can be convenient for this aim.

2 Impossibilities, theories and propositional content

Impossible worlds are especially important for the formalization of databases and scientific theories that are in general directly or indirectly inconsistent or incomplete.

As we have already seen classical logic does not preserve the content relation between the premises and the conclusion hence is not appropriate to explicate the natural scientific reasoning. But the most formal theories, which are based on empirical or experimental sciences, are often, as Cheng notes, indirectly inconsistent, i.e., they admit the truth of some statement and its negation and there is no evidence for deciding which of them is true; on the other hand a scientific theory may be incomplete in many ways, so a contention as well as its negation fail to be true in the theory. Thus the reasoning in the first case has to be paraconsistent in order to avoid the necessity to infer (to think) an arbitrary statement from any contradiction and respectively the reasoning in the second case has to be incomplete in order to avoid the necessity to think about things which are out of the theory (Cheng, 2006, p. 311). The situation in databases is similar — see (Restall, 1999, p. 69), (Varzi, 1997, p. 625).

Impossible worlds have also a useful role about propositional content. The latter is in a close relation with the intensional context which content may be inconsistent. The propositional attitudes with impossible content such as *I tie each of my shoes before the other* or *Fermat's Last Theorem is false* and etc. (Barwise, 1997, p. 490) in the light of possible world semantics correspond to the empty set of worlds and although they have different content it cannot be distinguished. The usage of impossible worlds however gives them a chance to present in the logical transformations and the contextual difference between them to be analyzed. It happens because in the semantics of impossible worlds two impossible pairs are not the same thing.

Since in their practical activity people often consider what would be the case in various impossible situations, the usage of impossible worlds is also reasonable in respect to pair counterfactuals with the same metaphysically impossible antecedents and different consequents. Mares uses the next example:

(1) *If Sally were to square the circle, we would be surprised* and

(2) *If Sally were to square the circle, we would not be surprised,*

where (1) seems to us true while (2) seems false (Mares, 2004, p. 516). In order to deal systematically and non-trivially with them and to be shown their intensional difference are needed tools distinguishing these two kinds of statements. It cannot happen without impossible worlds, because in the possible ones the contradiction is always false, which means that the consequent has no matter and two counterfactuals of this kind cannot be distinguished.

Except the discussed kind of impossibilities by imposing different worlds, there is other kind of impossibilities — by imposing different objects. In the second case the impossibility arises when are taken two different objects which properties are inconsistent one another and with such objects is formed a new one (Restall, 1997, p. 593–594). Instances of such kind are: circle square, square circle, being green and colorless and so on (Vander Laan, 1997, p. 599).

This kind of impossibilities can be justified epistemologically (according to Barwise), if we are not concerned by the metaphysical possibility, but by the epistemological one — if the agents have known about the circle and the square only how they can arise from the surface (Barwise, 1997, p. 499).

I constructed the following example in order to make clearer in which cases the beliefs in such impossibilities are justifiable (see the picture on page 11).

In this case the only thing the agent knows is that it may be constructed a square from (1) and a circle also from (1). Looking at the sketches (2) and (3) we see that it is really possible. So, the only information which the agent has about the circle and the square is that they can be constructed from two perpendicular lines as is shown on the sketch (1). He/she has no other knowledge about the properties

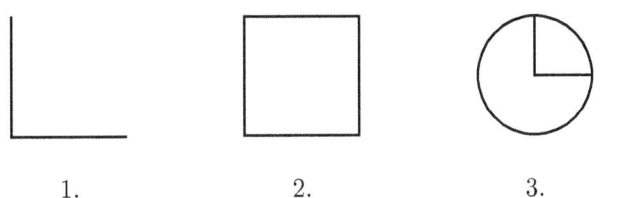

1. 2. 3.

of these figures. Because of that he/she supposes that it is possible to be constructed a figure having the properties of both the circle and the square (as they can be constructed from the same components), namely — a circle square. Therefore a notion for such kind of impossible objects can be formed from agents who are beginners in a certain sphere or who are deprived from considerable information.

The intensional contexts include also the perception. Illustrations about our ability to perceive impossibilities are for example the drawings of Escher. For the explanation of such "phenomena" are used by Mortensen and Penrose the theory of cohomology groups and the theory of heap (Mortensen, 1997, p. 532).

3 Inconsistent and incomplete worlds and the problem of vagueness

Actually inconsistent and incomplete models have also other applications, namely in regard to some theories of vagueness (I mean supervaluationism and subvaluationism) as far as they are characterized with gaps about truth-values and respectively with gluts concerning truth-values. Briefly, as Hyde shows, since for vague predicates there is not any sharp boundary that could conceivably be drawn separating the predicate's positive extension from its negative extension, on the view of supervaluationism, where a is a borderline case of F, the indeterminacy of Fa amounts to its being neither true nor false (Hyde, 2008, p. 73–74), so borderline cases are giving rise to truth-value gaps. On the other hand subvaluational semantics treats borderline cases for a vague predicate like "heap", namely as cases to which the predicate both applies and does not apply. That is, if a is a borderline case for "heap" then "a is a heap" is true and "a is not a heap" is true, hence the indeterminate (vague) sentences take on both truth-values giving rise to truth-value gluts (Hyde, 2008, p. 94–95). These gaps and gluts ensure the non-triviality of incompleteness (i.e. $B \not\vDash A, \sim A$), respectively the non-triviality of inconsistency (i.e. $A, \sim A \not\vDash B$) and are also used with respect to sorites paradoxes: the former — to falsify some of the premises, and the last — to invalidate modus ponens (or the disjunctive syllogism). Thereby, these inconsistent and incomplete models are applicable in the mentioned theories, although they are not exactly the same models as these ones which were regarded in the previous sections, because the logics which govern the intuitions of subvaluationism and supervaluationism are only weakly paraconsistent and respectively weakly paracomplete, i.e. although the failure of bivalence where the first theory admits A and $\sim A$ to be true — it retains the law of non-contradiction (and of excluded middle) as well as in the second theory although both A and $\sim A$ may fail to be true it retains the law of excluded middle (and of non-contradiction as well). These circumstances make conjunction and respectively disjunction non-truth functional and raise a lot of difficulties which I will not consider here. Anyway, I aimed to outline the application of the mentioned models in this sphere although, in agreement with Hyde, I think that the strong paraconsistent or paracomplete response may contribute more to the analyses of vagueness and will be closer to our intuitions about vague predicates and propositions.

There is also another logic concerning the problem of vagueness where the inconsistent and incomplete worlds present. Priest, for example, constructs fuzzy relevant logic inspired on the one hand from vagueness and on the other hand — from relevance. Actually he uses two approaches to construct such logic — the first one is as fuzzifying relevant world semantics (Priest, 2002, p. 263) and the second one — reinterpreting the algebraic semantics for relevant logic (Priest, 2002, p. 273). I will mention a few words for the first approach as far as it includes the discussed worlds as a part of the relational semantics of relevant logic. In broad outlines Priest transforms the relevant semantics into a fuzzy logic, simply replacing the set of truth values $\{0, 1\}$ with the closed interval $[0, 1]$ (Priest, 2002, p. 266). The interpretation is augmented with the operator $*$ (the Routley star) on worlds, as well as with the ternary relation R on worlds (Priest, 2002, p. 267–269), but without any suggestions concerning the meaning of both of them. The last is a warrant for me to claim that fuzzy relevant logic presented in this variant doesn't help a lot for the clarification of the problem of vagueness (although it is inspired of it) because it doesn't give to the standard fuzzy logic new interpretation and new semantic value of vague propositions (for the last aim is used the standard for fuzzy logic continuum of truth values) and doesn't exemplify any new characteristics of vagueness. Maybe the reason is due to the circumstance that actually the specific for relevant logic α and α^* worlds which could be used in order to referee to any vague propositions and to be interpreted eventually as any worlds of vagueness are not used in this way — they are included here together with the ternary relation as technical means ensuring the non-triviality of inference. It seems to me that the proposed variant makes fuzzy logic more relevant than giving solutions to the problem of vagueness. The last, of course, is a sufficient condition to be scrutinized this logic and to be made more investigations concerning its principles. On the other hand Priest doesn't show how to treat the sorites paradox by this apparatus. Since modus ponens for material implication (actually the disjunctive syllogism) is not valid here which is not a surprise as it is not valid in relevant logic, maybe it could be used in harmony with the arguments from subvaluationism in regard to this issue. Thus, although fuzzy relevant logic includes inconsistent and incomplete worlds the last don't find sufficient applications concerning vagueness which doesn't mean that any future inquiries will not contribute to this area and they (with any additional constraints or any changes in the interpretation) couldn't be helpful in the discussed problem field, not only for the non-triviality of inference as is here. Maybe is justifiable to look at the relational semantics of relevant logic and especially at α and α^* worlds in the light of vagueness — we have some warrants about that.

4 Conclusion

The nature of impossible worlds could be: ersatz constructions (Lewis);[3] elements of any set-theoretical model for which is stipulated to be under certain constraints (Mares, Restall, Varzi);[4] abstract objects — which existence is inferred or abstracted in rational way (Zalta, Vander Laan, Barwise)[5] and etc., which depends on the corresponding philosophical position and respectively enhance their theoretical significance. But no matter what they are, inconsistent

---

[3] See (Mares, 2004, pp. 517, 520).
[4] See (Mares, 2004), (Restall, 1997), (Varzi, 1997).
[5] See (Barwise, 1997, p. 491), (Zalta, 1997), (Vander Laan, 1997).

and incomplete worlds, as we saw, contribute to the clarification of logical validity and make a comprehensive analysis of negation (Beall & Restall, 2000, pp. 481–482).

They also permit to be done operations with logically undetermined concepts or when there is a lack of information on some issue especially in different epistemic situations. Impossible worlds afford an opportunity to include the inconsistencies in our reasoning and to work with them until precise and revise the available information and differentiate exactly our contentions. On the other hand they give resources to be excluded from our logical transformations those statements which are indifferent for our system. In a combination with some additional requirements they may be helpful to the theories of vagueness.

Doroteya Angelova
Institute for Philosophical Research
Bulgarian Academy of Sciences
6, "Patriarch Evthimii" blvd.
Sofia 1000, Bulgaria
doroteyaan@gmail.com; teiaang@yahoo.com
http://www.philosophybulgaria.org/en/Sekcii/Logika/Sastav.php

References

Anderson, A., & Belnap, N. (1975). *Entailment. The Logic of Relevance and Necessity* (Vol. I). Princeton: Princeton University Press.

Angelova, D. (2004). Relational Semantics of Relevant Logic. *Philosophical Alternatives*, *4–5*, 196–206. (in Bulgarian)

Barwise, J. (1997). Information and Impossibilities. *Notre Dame Journal of Formal Logic*, *38*(4), 488–515.

Beall, J. C., & Restall, G. (2000). Logical Pluralism. *Australasian Journal of Philosophy*, *78*(4), 475–493.

Beall, J. C., & Restall, G. (2006). *Logical Pluralism*. Oxford: Oxford University Press.

Cheng, J. (2006). Strong Relevant Logic as the Universal Basis of Various Applied Logics for Knowledge Representation and Reasoning. In Y. Kiyoki, J. Henno, H. Jaakkola, & H. Kangassalo (Eds.), *Information Modelling and Knowledge Bases XVII* (p. 310–320). IOS Press. (Frontiers in Artificial Intelligence and Applications, vol. 136)

Hyde, D. (2008). *Vagueness, Logic and Ontology*. Burlington, VT.: Ashgate.

Mares, E. (1997). Who's afraid of impossible worlds. *Notre Dame Journal of Formal Logic*, *38*(4), 516–526.

Mares, E., & Meyer, R. (2001). Relevant Logics. In L. Goble (Ed.), *The Blackwell Guide to Philosophical Logic* (p. 280–308). Oxford: Blackwell Publishers Ltd.

Mortensen, C. (1997). Peeking at the Impossible. *Notre Dame Journal of Formal Logic*, *38*(4), 527–534.

Perry, J. (1998). Semantics, Situation. In E. Craig (Ed.), *Routledge Encyclopedia of Philosophy, vol. 8 (Questions — Sociobiology)* (p. 669–672). London: Routledge.

Priest, G. (1997). Editor's Introduction. *Notre Dame Journal of Formal Logic*, *38*(4), 481–487.

Priest, G. (2002). Fuzzy Relevant Logic. In W. Carnielli, M. Coniglio, & L. D'Ottaviano (Eds.), *Paraconsistency. The Logical Way To The Inconsistent* (p. 261–274). (Proceedings of the World Congress Held in Sao Paulo)

Restall, G. (1997). Ways Things Can't Be. *Notre Dame Journal of Formal Logic*, *38*(4), 583–596.

Restall, G. (1999). Negation in relevant logics: How I stopped worrying and learned to love the Routley star. In D. Gabbay & H. Wansing (Eds.), *What is Negation?* (p. 53–76). Dordrecht: Kluwer. (Applied Logic Series, vol. 13)

Vander Laan, D. (1997). The Ontology of Impossible Worlds. *Notre Dame Journal of Formal Logic*, *38*(4), 597–620.

Varzi, A. (1997). Inconsistency without contradiction. *Notre Dame Journal of Formal Logic*, *38*(4), 621–639.

Zalta, E. (1997). A Classically Based Theory of Impossible Worlds. *Notre Dame Journal of Formal Logic*, *38*(4), 640–660.

On truth, abnormal worlds, and necessity

Jc Beall*

1 Introduction

Various semantic theories (e.g., truth, exemplification, and more) are underwritten by so-called depth-relevant logics. Such logics afford non-trivial theories that enjoy unrestricted semantic principles (e.g., T-biconditionals, comprehension, etc.). Standard semantics for such logics are so-called non-normal-worlds semantics, which add 'abnormal worlds' (or 'non-normal worlds') to an otherwise standard possible-worlds framework. (All of these ideas are briefly reviewed below.)

Once worlds (of any sort) are in the picture, questions about other worlds-involving notions emerge. One issue concerns the addition of standard alethic modalities — e.g., necessity — into the picture. In this paper, I note that the addition of such modalities (e.g., necessity, on which I focus here) is not entirely straightforward. In particular, the problems that motivate the target (depth-relevant) semantic theories — namely, Curry-paradoxical problems — equally constrain the treatment of alethic modalities.[1]

The paper runs as follows. §§ 2–3 review Curry's paradox and its upshot for target semantic theories. § 4 sketches the target (abnormal-worlds) semantics as background to the main issue. The main issue is

---

*I am very grateful for the opportunity to have spoken at LOGICA2009. I not only learned a great deal from the variety of logically relevant talks; I also received valuable feedback on my paper (which you are now reading) and also on a more general result discussed in a different (to-appear) paper. I particularly thank the organizers of LOGICA2009, whose generosity made my experience nothing but enjoyable — or, at least, as with the walk through a 'river' on a rainy day, very memorable! Versions (and variations) of this paper also benefited from audiences at the 2009 Australasian Association for Philosophy and Australasian Association for Logic conferences in Melbourne, Victoria University of Wellington, Auckland University, and the University of Otago. Thanks to all who gave useful feedback at some stage or other, but particularly to Greg Restall, Graham Priest, Steve Read, Aaron Cotnoir, Dave Ripley, Ed Mares, Denis Robinson, Max Cresswell, Josh Parsons, and Charles Pigden.

[1] I concentrate solely on necessity in this paper. A more general result concerning other modalities is discussed in a separate paper (Beall, 2009a).

discussed in §5 and §6. A solution to the target problem is given in §7, with §8 giving a few closing remarks.

2 Robust contraction freedom

A lesson commonly drawn from Curry's paradox is that semantic principles (e.g., truth biconditionals, exemplification or comprehension biconditionals, etc.) need to be free from various forms of contraction (Meyer, Routley, & Dunn, 1979; Priest, 2006b; Field, 2008; Beall, 2009b). A simple way of seeing the point is via a result of Greg Restall's (1993a).

Let $\to$ be a rule-detachable conditional (our target detachable arrow),[2] and let $\odot$ be a binary connective. Following Restall (1993a), we call $\odot$ a *contracting connective* if all of the following hold.[3]

$$A \to B \vdash A \odot B \qquad (1)$$

$$A \odot (A \odot B) \vdash A \odot B \qquad (2)$$

$$A, A \odot B \vdash B \qquad (3)$$

For Curry-paradoxical reasons, any contracting connective trivializes a (sufficiently expressive) language that enjoys a truth predicate for which all of the target-arrow biconditionals hold (where $\leftrightarrow$ is formed via our target detachable arrow and conjunction in the usual way):[4]

$$\text{Tr}(\langle A \rangle) \leftrightarrow A \qquad (4)$$

To see this, let $\odot$ be a contracting connective, and $\bot$ an explosive sentence (i.e., implies all sentences), and let k name the sentence

---

[2]By *rule-detachable* is meant that, according to the given logic (or consequence relation $\vdash$), A and $A \to B$ jointly imply B, that is, that the argument from from A and $A \to B$ to B is valid (according to the given logic). Henceforth, I use 'detachable' just to mean *rule-detachable*.

I should also note, with respect to notation, that I use the turnstile (throughout) to record the validity relation; it may be read as *implies*, so that '$A \vdash B$' amounts to the claim that A implies B or — equivalently — that B is a consequence of A.

[3]Restall calls it a 'contracting implication', but I will avoid this terminology. (Some folks might worry about whether it's really an *implication* or even a *conditional* or etc., but this is irrelevant to the current discussion, and so I simply sidestep by using the generic 'connective'.)

[4]Think of $\langle A \rangle$ as a structural-descriptive name of A (or, if you wish, Gödel codes, or some such suitable naming device). Also, we assume that conjunction is standard (obeying both Adjunction and Simplification), that is, $A, B \vdash A \wedge B$ and $A \wedge B \vdash A$ (similarly for B).

$\mathrm{Tr}(k) \odot \bot$, so that we have the following true identity statement:
$$k = \langle \mathrm{Tr}(k) \odot \bot \rangle \tag{5}$$

The target (Curry-) instance of (4) is
$$\mathrm{Tr}(\langle \mathrm{Tr}(k) \odot \bot \rangle) \leftrightarrow \mathrm{Tr}(k) \odot \bot \tag{6}$$

From this and our identity (5) we get
$$\mathrm{Tr}(k) \leftrightarrow \mathrm{Tr}(k) \odot \bot \tag{7}$$

Applying (1) to the LRD of (7) yields
$$\mathrm{Tr}(k) \odot (\mathrm{Tr}(k) \odot \bot) \tag{8}$$

Applying (2), the basic contraction rule, to (8) yields
$$\mathrm{Tr}(k) \odot \bot \tag{9}$$

But, now, (9) and the RLD of (7) deliver
$$\mathrm{Tr}(k) \tag{10}$$

The final blow comes from (3), which, applied to (9) and (10), delivers $\bot$, which implies all sentences. Triviality — everyone's worst nightmare.

3 The upshot

For convenience, let us call the conditional involved in one's semantic principles (e.g., truth biconditionals, exemplification or naïve-membership comprehension, and so on) a *semantic conditional*. Let us assume, as I will throughout, that our semantic conditional is a *detachable* conditional.[5]

An immediate upshot of § 2 is that, on pain of its being a contracting connective, one's semantic conditional — say, the conditional in the truth biconditionals — cannot satisfy (2). (Letting $\odot$ be $\rightarrow$ in (1)

---

[5] I myself endorse such a position in (2009b), as do Priest (2006b, 2006a), Field (2008), and many other theorists. I should note, however, that I have lately come to seriously question this assumption, but discussion of that topic is for another work (Beall, 20++).

and (3) makes the point plain.) The more general point is that having a non-contracting semantic conditional is not itself sufficient to avoid Curry problems. What one cannot have is *any* contracting connective in the language, lest the resulting semantic theory be trivial (via the considerations in §2). Following Restall (1993a), we say that a language (or theory in the given language) is *robustly contraction-free* just if it is free of a contracting connective.

4 Abnormal worlds and depth-relevant theories

There are a variety of logics that provide robustly contraction-free semantic theories. My concern here is a family of so-called depth-relevant logics, which have been used to provide non-trivial semantic theories (Brady, 1989; Priest, 2006a, 2006b; Beall, 2009b).[6]

The target logics enjoy a possible-worlds semantics called *abnormal-worlds semantics* or, more commonly, *non-normal-worlds semantics*.[7] The point of this section is not to sketch the full semantic framework(s) in question, but rather to sketch just enough of the target framework to raise the main issue of the paper (concerning the addition of necessity into the mix).

** *Parenthetical remark.* I should note that my real target semantics are so-called simplified Routley–Meyer semantics that involve a ternary relation on worlds (Routley & Meyer, 1973). It is this ternary relation that is invoked to achieve a detachable but contraction-free semantic conditional in some of the target semantic theories (Priest, 2006b; Beall, 2009b), and in many ways is at the heart of (at least worlds) semantics for relevant logics (Priest & Sylvan, 1992; Restall, 1993b). But simply for simplicity here, I present a different, 'arbitrary-evaluator' framework (Routley & Loparić, 1978; Priest, 1992; Beall, 2005). *End parenthetical.* **

---

[6]Though not depth-relevant, the recent theories of Field (2008) and Brady (2006) fall into the same target family of semantics theories; however, there are serious differences from the target depth-relevant theories, and I do not discuss the Brady/Field (or their ilk) theories further here.

[7]Kripke (1965) first invoked non-normal worlds to model weak C. I. Lewis modal systems, but they're now useful for other things — epistemic logics, and, on topic here, for conditionals.

4.1 Abnormal-worlds structures

Although many semantic paradoxes arise only at the predicate–quantifier level, the main ideas of this paper can be conveyed at the propositional level. So, for simplicity, we concentrate only on a simple (positive) propositional language with $\wedge$, $\vee$, and $\rightarrow$. (As mentioned above, we can ignore negation. In fact, we can even ignore conjunction and disjunction for present purposes, but I sketch their treatment for purposes of comparison with the 'jumpy' treatment of the arrow.)

Our (abnormal-worlds) interpretations are much like standard possible-worlds interpretations except for an additional non-empty set $\mathcal{N}$ comprising the *normal worlds* of the interpretations. Abnormal-worlds structures are pairs $\langle \mathcal{W}, \mathcal{N} \rangle$, where $\mathcal{W}$ is a non-empty set of worlds (more neutrally, points) and $\mathcal{N}$ a non-empty subset of $\mathcal{W}$. If $x \in \mathcal{N}$ we call x a *normal world* (or normal point), and if $x \in \mathcal{W} \setminus \mathcal{N}$ we call x an *abnormal* world. (NB: $\mathcal{W} \setminus \mathcal{N}$ may be empty.)

We let $\models$ be a truth-at-a-point relation, relating sentences to worlds. In particular, $\models$ may be any subset of $\mathcal{W} \times \mathcal{L}$ (where $\mathcal{L}$ comprises our sentences). If $x \models A$ we say that A is (at least) true at x.[8]

For present purposes, we say that any abnormal-worlds structure, combined with a truth-at-a-point relation, is an *abnormal-worlds interpretation*. (We specify 'admissible interpretations' or models below.)

4.2 Models: truth conditions

A characteristic feature of abnormal-worlds semantics is that the truth conditions for connectives may be 'jumpy' or 'non-uniform' across types of points: a connective might behave differently — have different truth-at-a-point conditions — at different types of points. For convenient terminology, we say that a connective is *jumpy* or *non-uniform* iff its truth-at-a-point conditions vary across normal and abnormal worlds; otherwise, we call the connective *uniform*. (The distinction will be clear from the truth conditions for the conditional below.)

---

[8] The parenthetical 'at least' flags that the target theories are paraconsistent, allowing sentences to be true and also false at a point. For present purposes, however, we can ignore this issue — focusing only on a *truth-at-a-world* relation (ignoring falsity-at-a-point).

Models. In the target semantics, conjunction and disjunction are uniform, but the target 'semantic conditional' is jumpy. In particular, we call any abnormal-worlds interpretation a *model* — or, if you like, *admissible interpretation* — if and only if it 'obeys' the following 'truth conditions' for our connectives. (For present purposes, we can ignore 'falsity conditions' for our connectives.)

1. Conjunction:
   - Normal or Abnormal: for any $x \in \mathcal{W}$

   $$x \models A \wedge B \text{ iff } x \models A \text{ and } x \models B$$

2. Disjunction:
   - Normal or Abnormal: for any $x \in \mathcal{W}$

   $$x \models A \vee B \text{ iff } x \models A \text{ or } x \models B$$

3. Conditional:
   - Normal: for any $x \in \mathcal{N}$

   $$x \models A \rightarrow B \text{ iff } y \not\models A \text{ or } y \models B \text{ for } any\ y \in \mathcal{W}$$

   - Abnormal: for any $x \in \mathcal{W} \setminus \mathcal{N}$

   $$x \models A \rightarrow B \text{ iff } \ldots \text{let this be arbitrary!}$$

The foregoing truth conditions are familiar except, perhaps, for the conditional's truth-at-a-point conditions at *abnormal worlds*. At abnormal worlds, the truth (or semantic status, generally) of $A \rightarrow B$ is entirely arbitrary.[9]

Validity. We define *validity* only over normal worlds (of all models): the argument from $A_1, \ldots, A_n$ to B is valid iff there's no *normal* world of any model at which each A_i is true but B not.

---

[9] I should repeat that, in the target logics, one need not make the conditional's status arbitrary at abnormal worlds, but this approach simplifies matters for current discussion. For discussion of this 'arbitrary' approach, see Priest, 1992 and Beall, 2005.

4.3 Example

Note that the foregoing framework delivers a detachable but non-contracting conditional. Letting $\vdash$ be our validity relation, we have detachment:
$$A, A \to B \vdash B.$$
Suppose, for reductio, that there's a point $x \in \mathcal{N}$ at which A and $A \to B$ are true but B untrue (i.e., $x \not\models B$). Since, by supposition, $x \models A \to B$ and $x \in \mathcal{N}$, there's no point $y \in \mathcal{W}$ such that $y \models A$ and $y \not\models B$, and a fortiori either $x \not\models A$ or $x \models B$. Contradiction.

With respect to contraction, we have
$$A \to (A \to B) \not\vdash A \to B.$$
In particular, let $\mathcal{W} = \{x, y\}$ with $\mathcal{N} = \{x\}$. Let $x \not\models A$. In turn, let $y \models A$ and $y \not\models B$ but $y \models A \to B$. In this model, there's no point in which A is true but $A \to B$ untrue, and hence, since x is normal, we have $x \models A \to (A \to B)$. On the other hand, there is a point at which A is true but B untrue (viz., y, which is abnormal); and so $x \not\models A \to B$.

5 Target issue: necessity

While there are a host of philosophical questions that surround the given abnormal-worlds semantics, my interest here is in a relatively unexplored one: namely, the addition of necessity into the mix. In what follows, I note that, on pain of generating a contracting connective (and, hence, trivializing the target theories), a necessity operator along standard *quantifier-over-all-worlds* lines must be 'jumpy' or non-uniform.

5.1 Uniform, all worlds

Let us assume that we've added a unary connective $\Box$ to be our (broad) necessity operator (viz., *it is necessary that...*). One natural thought for truth conditions is a uniform, all-worlds condition:

- Normal or Abnormal: for any $x \in \mathcal{W}$
$$x \models \Box A \text{ iff } y \models A \text{ for all } y \in \mathcal{W}.$$

This is the standard idea prevalent in philosophy: namely, that our broad-necessity operator ranges over *all* worlds, and (for 'uniformity') does as much at *all* points. (If we think in terms of an 'access relation' on worlds, then the idea is that we have an equivalence relation on worlds.) The problem with this is that, not surprisingly (Kripke, 1965), we can now (i.e., in our *abnormal*-worlds framework) have logical truths that do not count as necessarily true. For example, one may easily check that $A \to A$ is logically true (i.e., no countermodel), but there are many worlds (of many models) at which $A \to A$ is untrue — namely, abnormal worlds. At least on the surface, this break between alethic necessity and logical truth seems awkward. On the other hand, perhaps the given break between aletheic necessity and logical truth — more formally, a failure of Necessitation — is not only what one should expect in the current (abnormal-worlds) conext; it's what one should want, since abnormal-worlds are strange (e.g., $A \to A$ can fail to be true!), so strange that we define validity only over the normal worlds. I will not argue the matter here.

We can allow (if we want) that our broad notion of necessity ought to be as broad as the universe of points that we recognize, and that if (as we're supposing) we recognize abnormal points, then our broad-necessity operator ought to range (i.e., quantify over) them too. Still, one might (quite reasonably) think that we have another notion of necessity, namely, one that quantifies only over *normal* worlds — the 'real possibilities', so to speak. This is the notion of necessity at issue in this paper.

5.2 Uniform, all *normal* worlds

While abnormal worlds might be strange enough to be 'possibilities' only in some very charitable sense, it is natural to take the *normal worlds* to be the 'real possibilities' involved in our alethic-necessity (or, dually, possibility) claims.

- Normal or Abnormal: for any $x \in \mathcal{W}$

  $$x \models \Box A \text{ iff } y \models A \text{ for all } y \in \mathcal{N}.$$

In short: we take our necessity claims to quantify over all and only the *normal* worlds. Not only does this deliver the necessity of all logical truths; it also accommodates the general intuition that the abnormal worlds are 'impossible worlds' of some sort (Priest, 1992; Caret, 2009), or at any rate that abnormal worlds are beyond the intended range of our necessity claims.

It is not difficult to see that we also get a lot of standard logical behavior for the Box. One notable feature is the following lemma (not unfamiliar from the more standard S5 setting).

Lemma 1 (Box Lemma). *For any model, $\Box A$ is true at all worlds or true at none.*

Proof. This is fairly clear from the truth conditions for the box. In short: either A is true at all *normal* worlds or not. If the former, $\Box A$ is true everywhere (by the given truth conditions). If the latter, $\Box A$ is true nowhere (by the given truth conditions).

Of course, whether one gets standard S5 interaction between the Box and Diamond depends on how negation is treated, at least if we're defining the Diamond in terms of negation and the Box along standard lines. But this issue is beyond the limited aims of this paper.

6 Trouble: contracting connective

However natural the uniform-all-normal-worlds account of §5.2 may be, it cannot be utilized in the target semantic theories. The trouble, in short, is that the account breeds a contracting connective.

To see the result, let us define a connective $\Rightarrow$ as follows:

$$A \Rightarrow B := \Box(A \to B).$$

Our defined connective is a contracting connective. To see this, first note a general lemma concerning the arrow.

Lemma 2 (Arrow Lemma). *For any model, $A \to B$ is true at some normal world if and only if it is true at all normal worlds.*

Proof. This is fairly clear from the truth-at-a-*normal*-point conditions for the arrow: $A \to B$ is true at a normal world if and only if there's *no point whatsover* at which A is true and B untrue. So, $A \to B$ is true at *some* normal point iff true at *all* normal points. (Think about the Box in an S5 setting. The situation is the same here when we restrict to normal worlds.)

The conditions for contracting connectives are now met as follows.

1. $A \to B \vdash A \Rightarrow B$.

    Proof. This follows from the Arrow Lemma and the truth conditions for the Box. A countermodel would require a normal world at which $A \to B$ is true but also (to make the conclusion untrue) a *normal* world at which $A \to B$ is untrue. This contradicts the Arrow Lemma.

2. $A \Rightarrow (A \Rightarrow B) \vdash A \Rightarrow B$.

    Proof. Suppose that $x \in \mathcal{N}$ and $x \models \Box(A \to \Box(A \to B))$, in which case, via the Box's truth conditions, $x \models A \to \Box(A \to B)$. Suppose, for reductio, that $x \not\models \Box(A \to B)$, in which case there's some $y \in \mathcal{N}$ such that $y \not\models A \to B$, and so — via the Arrow's normal-point conditions — some $z \in \mathcal{W}$ such that $z \models A$ and $z \not\models B$. Now, given that $x \not\models \Box(A \to B)$, the Box Lemma implies that $z \not\models \Box(A \to B)$. But, then, there's a point (viz., z) at which A is true but $\Box(A \to B)$ untrue — a point at which A is true but $A \Rightarrow B$ untrue. Hence, by the arrow's normal-point conditions, $x \not\models A \to (A \Rightarrow B)$. Contradiction (see initial supposition).

3. $A, A \Rightarrow B \vdash B$.

    Proof. Let $x \in \mathcal{N}$ and $x \models A$ and $x \models A \Rightarrow B$. For reductio, suppose that $x \not\models B$, in which case $x \not\models A \to B$ since there's a point (viz., x) at which A is true but B untrue. But this contradicts the supposition that $x \models A \Rightarrow B$, which requires that $A \to B$ be true at all normal worlds — and, a fortiori, true at x.

The upshot: we cannot recognize an alethic-necessity operator that *uniformly* ranges over only normal worlds. What to do?

7 Solution: jumpy necessity

The solution — at least if (as I'm supposing) we want Necessitation to hold for our target necessity operator — is to give up on uniformity and treat our Box like we treat our arrow: namely, as a jumpy or non-uniform connective. And this makes sense. After all, we first invoked abnormal worlds to free our conditional from contraction —

freedom from feature (2) and its ilk. The conditional (i.e., our basic arrow) avoids contraction by going on holiday at abnormal worlds: it behaves differently — indeed, on the simple sketch here, arbitrarily — at abnormal worlds. (Look again at the countermodel to contraction in §4.3.) The trouble with our uniform, all-normal-worlds approach to the Box is that, because it is uniform, the arrow, when 'boxed up' (so to speak), is forced to behave normally; it is not allowed to be evaluated in its holiday state. In particular, while $A \to B$ can go on holiday at abnormal worlds, $\Box(A \to B)$ forces $A \to B$ to return (so to speak) to normal worlds for evaluation, and so our defined arrow $A \Rightarrow B$ never has a point at which to shake off contraction.

We can skip further metaphor and simply go to a solution. As above, the solution is to treat our Box as we treat the arrow (our other 'intensional' connective): treat it as a jumpy connective. There are many ways to do this, but the simplest (though, perhaps, philosophically ugliest) is to again invoke an 'arbitrary evaluator' at abnormal points. And to retain Necessitation (i.e., that if A is logically true, so too is $\Box A$), we retain the spirit of our all-normal-worlds condition at (and only at) normal worlds.

- Normal: for $x \in \mathcal{N}$

$$x \models \Box A \text{ iff } y \models A \text{ for all } y \in \mathcal{N}.$$

- Abnormal: for $x \in \mathcal{W} \setminus \mathcal{N}$

$$x \models \Box A \text{ iff } \dots \text{ let this be arbitrary!}$$

With this approach, we keep Necessitation.

- *Necessitation.* If $\vdash A$ then $\vdash \Box A$.

    Proof. This follows immediately from the fact that validity (even for sentences) is defined only over normal worlds: $\vdash A$ iff $x \models A$ for all $x \in \mathcal{N}$ in all models.

Moreover, and for present purposes more importantly, we avoid $A \Rightarrow B$'s being a contracting connective — where, again, $A \Rightarrow B$ is $\Box(A \to B)$.

- $A \Rightarrow (A \Rightarrow B) \nvdash A \Rightarrow B$.

    Countermodel. The same countermodel to (2) works here. Let $\mathcal{W} = \{x, y\}$ with $\mathcal{N} = \{x\}$. Let $x \nvDash A$. Now, let $y \vDash A$ and $y \nvDash B$ but $y \vDash \Box(A \rightarrow B)$. [Recall that box claims can be whatever we want at abnormal points.] In this model, there's no point at which A is true but $A \Rightarrow B$ untrue, and hence, since x is normal, we have $x \vDash A \rightarrow (A \Rightarrow B)$. Hence, since x is the sole normal world, we have $x \vDash \Box(A \rightarrow (A \Rightarrow B))$, that is, $x \vDash A \Rightarrow (A \Rightarrow B)$. On the other hand, there is a point (viz., y) at which A is true but B untrue; and so, by the Arrow's normal-point conditions, $x \nvDash A \rightarrow B$. By the Box's normal-point conditions, we have that $x \nvDash \Box(A \rightarrow B)$, that is, $x \nvDash A \Rightarrow B$.

While there may be other features of necessity that we may want, this *non-uniform* or 'jumpy', all-normal-worlds approach is at least promising.

** *Parenthetical remark.* Showing that $\Box(A \rightarrow B)$ fails to contract is insufficient for establishing robust contraction freedom, but for semantic theories in the ballpark (Priest, 2006b; Beall, 2009b), a non-triviality result is available. For example, such theories have one-normal-world models (with all other worlds abnormal). So, as Greg Restall (in conversation) noted, $\Box A$ is vacuously true in such models (i.e., true at the unique normal world of such models), and so we have models of the resulting Box-ful semantic theories (with the Box having our given 'jumpy' all-normal-worlds semantics). Of course, what one would like are *natural* models of such theories, where not *all* Box claims are true, and this — at the time of writing — remains an open problem. [The target sense of 'natural' is imprecise and relative to background philosophical issues concerning the target semantic notions, but an example of a *natural model* for a Box-free semantic theory, relative to a particular transparent conception of truth, is presented in my *Spandrels of Truth* (2009b), with the model essentially due to Ross Brady, Chris Mortensen (1995), and Graham Priest.] *End parenthetical.* **

8 Concluding remarks

Contraction-free logics remain popular (and, I think, promising) routes towards constructing rich semantic theories. Among such logics are depth-relevant logics, which enjoy a familiar (though abnormal-) worlds semantics. Little attention has been put to the issue of adding other philosophically important notions into the mix, probably because the addition of such operators seemed on the surface to be straightforward. (This is certainly why I gave the matter little thought until recently.) It is surprising that there is any issue at all. What this paper shows is that, while the solution (e.g., § 7) is straightforward, care must nonetheless be taken when adding otherwise very familiar (and philosophically important) intensional notions into the mix.

** *Parenthetical remark.* There is a more general result concerning 'talking about normal worlds' from which the present result about necessity (or other modal notions like Actuality, etc.) follows. I hope to publish the broader result, with more general discussion, in a larger paper (Beall, 2009a). *End parenthetical.* **

Jc Beall
Philosophy Department and UConn Logic Group
University of Connecticut
Storrs, CT 06269–2054 USA
jc.beall@uconn.edu
http://homepages.uconn.edu/$\sim$jcb02005
http://logic.uconn.edu

References

Beall, J. (20++). *Transparent truth, limited inconsistency, and detachment.* (To appear.)

Beall, J. (2005). Transparent disquotationalism. In J. Beall & B. Armour-Garb (Eds.), *Deflationism and paradox* (pp. 7–22). Oxford: Oxford University Press.

Beall, J. (2009a). *Contraction and normal worlds: an overspill result.* (To appear. Presented at the 2009 AAL in Melbourne, Victoria University in Wellington, and University of Otago in Dunedin.)

Beall, J. (2009b). *Spandrels of Truth.* Oxford: Oxford University Press.

Beall, J., Brady, R., Hazen, A., Priest, G., & Restall, G. (2006). Relevant restricted quantification. *Journal of Philosophical Logic*, *35*(6), 587–598.

Brady, R. T. (1989). The non-triviality of dialectical set theory. In G. Priest, R. Routley, & J. Norman (Eds.), *Paraconsistent logic: Essays on the inconsistent* (pp. 437–470). Philosophia Verlag.

Brady, R. T. (2006). *Universal logic*. Stanford: CSLI Publications.

Caret, C. (2009). *Non-normal, but not impossible*. (To appear. Available at http://sites.google.com/site/colincaret/workinprogress)

Field, H. (2008). *Saving Truth from Paradox*. Oxford: Oxford University Press.

Hájek, P., Paris, J., & Shepherdson, J. (2000). The liar paradox and fuzzy logic. *Journal of Symbolic Logic*, *65*, 339–346.

Kripke, S. (1965). Semantical analysis of modal logic II: Non-normal modal propositional calculi. In J. W. Addison, L. Henkin, & A. Tarski (Eds.), *The Theory of Models* (pp. 206–220). Amsterdam: North-Holland Publishing Co.

Meyer, R. K., Routley, R., & Dunn, J. M. (1979). Curry's paradox. *Analysis*, *39*, 124–128.

Mortensen, C. (1995). *Inconsistent mathematics*. Kluwer Academic Publishers.

Priest, G. (1992). What is a non-normal world? *Logique et Analyse*, *35*, 291–302.

Priest, G. (2006a). *Doubt Truth To Be A Liar*. Oxford: Oxford University Press.

Priest, G. (2006b). *In Contradiction* (Second ed.). Oxford: Oxford University Press.

Priest, G., & Sylvan, R. (1992). Simplified semantics for basic relevant logics. *Journal of Philosophical Logic*, *21*, 217–232.

Restall, G. (1992). Arithmetic and truth in Łukasiewicz's infinitely valued logic. *Logique et Analyse*, *139–140*, 303–312.

Restall, G. (1993a). How to be *Really* contraction-free. *Studia Logica*, *52*, 381–391.

Restall, G. (1993b). Simplified semantics for relevant logics (and some of their rivals). *Journal of Philosophical Logic*, *22*, 481–511.

Routley, R., & Loparić, A. (1978). Semantical analyses of Arruda-da Costa *P*-systems and adjacent non-replacement systems. *Studia Logica*, *37*, 301–320.

Routley, R., & Meyer, R. K. (1973). Semantics of entailment. In H. Leblanc (Ed.), *Truth, Syntax, and Modality* (pp. 194–243). North Holland. (Proceedings of the Temple University Conference on Alternative Semantics)

From Achievement Stit to Metric Possible Choices

Roberto Ciuni*

1 Introduction

Stit logic is an established paradigm in the logic of agency, named after the acronym of "seeing to it that" (from now on, "stit"). Such a locution is read as a modal operator, and its truth-conditions are given on the basis of branching-time structures where, at each moment, histories are gathered into disjoint sets called *possible choices*. The formulae expressing the idea of an agent seeing to something are called "agentives", while the formulae bound by a stit operator are called "complements". There are two salient kinds of stit operator: the *deliberative* one and the *achievement* one. The former expresses the idea that something happens due to the deliberation of an agent (see (Horty & Belnap, 1995) and (Belnap, Perloff, & Xu, 2001))[1]. The latter expresses the idea that, because of a previous choice made by the agent, the world has been determined to lie in some of many possible developments of the events (see (Belnap & Perloff, 1992) and (Belnap et al., 2001), what "determining" exactly means here will be clear in section 2). Here, I shall focus on the achievement stit operator. The reason is that achievement stit captures the temporally extended character of "doing something" in a way the deliberative stit cannot. Indeed, in order to establish the truth-value of an agentive that contains the achievement stit, one needs two moments: a moment where the choice is made (usually called "witness") and a *later* moment where the action is accomplished (which I call "accomplishment") and where the agentive is to be evaluated. As a consequence,

---

*This work was carried out while I was a post-doctoral researcher at the Delft University of Technology, Section of Philosophy, with the project Applied Modal Logic.
I'd like to thank Nuel Belnap, Heinrich Wansing and Greg Restall for their fruitful comments.

[1]Such a notion is connected to Richmond Thomason's notion of deliberative obligations and traces back to Aristotle's analysis of deliberation in Nycomachean Ethics.

the achievement stit is implicitly influenced by the "points of view" of both moments: the one where a choice is made, and the one where an action is accomplished. By constrast, the deliberative stit considers just the moment where a choice is made.[2] In addition, I shall confine myself to the mono-agent case. That is to say, I shall use structures and languages where just one agent is taken into account.

As we shall see below (section 2), the classical achievement stit operator (introduced in (Belnap, 1991), (Belnap et al., 2001), and (Horty & Belnap, 1995)) does not verify Axiom K. In other words, $(astit_a(A \to B) \land astit_a A) \to astit_a B$ is not valid. This is an *indesideratum*, since the logics with axiom K are formally more well behaved than those without the axiom. In the present paper, I propose to define the achievement stit operator **n**-$astit_a$, and I prove that **n**-$astit_a$ makes Axiom K valid (section 3). **n**-$astit_a$ is a *metric achievement stit operator*, that is an achievement stit operator that is able to specify the distance between one intended moment of choice and the moment where the action is accomplished. In order to define **n**-$astit_a$, I shall introduce the tempo-modal language **MPC** which (1) is *metric*, this meaning that it is capable of expressing the idea of a distance between moments; (2) has operators that restrictedly quantify over histories in *a given possible choice* and one operator that quantifies over possible choices (the latter being a key notion in the semantics of any stit operator). **MPC** is inspired by the metric languages introduced by (Øhrstøm & Hasle, 1995), (Øhrstøm, 2009), and by the quantification over classes of histories presented in (Zanardo, 1998), (Zanardo, 2009). The introduction of **MPC** results in more than in a merely technical advantage. First, **MPC** allows us to express a conceptual insight that is very important for Stit logic: the idea that we can have some form of *determination without inevitability* (section 4). Such an insight cannot be expressed in the language for Stit logic or in classical languages for branching time. Second, **MPC** helps understanding how the achievement stit can be defined in terms of tempo-modal operators, and this in turns raises the question: are *agentives* somehow *dispensable* from Stit logic? The paper is divided as follows: in section

---

[2]For the deliberative stit, see (Belnap et al., 2001), pp. 29-38 and Chapters 11, 12 and 17, and (Horty & Belnap, 1995). The stit operator introduced by Chellas is very close to the idea behind deliberative stit. See (Chellas, 1992), (Belnap et al., 2001) (p. 248) and (Horty & Belnap, 1995) (pp. 599-600).

astit and **MPC**   35

2, I introduce $astit_a$ and its semantics, and I prove that $astit_a$ does not make Axiom K valid. In section 3, I introduce **MPC**, I define n-$astit_a$ and I prove it makes Axiom K valid. Section 4 sums up the proposal and discusses the aspects of **MPC** that are philosophically interesting.

Here, all the references to Stit logic and the *stit* operator must be understood in relation to the works of Belnap et al. that I am going to mention.

2   Stit Logic

In what follows, I shall read $astit_a$ A simply as "agent *a sees to it that A*", since the absence of a deliberative stit operator makes it superfluous to makes it clear that the operator expresses achievement stit. Let me define the stit language $\mathbf{AS} := \mathbf{PL} + \{astit_a\}$, where **PL** is the language of classical logic. $p_1, p_2, \ldots$ stay for *atoms*, $A, B, \ldots$ stay for arbitrary *formulae*. The usual symbols for the boolean connectives are employed. The formulae in **AS** are evaluated on the basis of stit-trees of the form

$$\mathcal{T}^{stit} := \langle \mathcal{T}, J, Ag, Choice \rangle, \qquad \text{where}$$

- $\mathcal{T}$ is a pair $\langle T, < \rangle$ with:

  1. T an infinite set of moments (t, t', t'', ...). Beside infinity, I make the assumption that T is not *more than continuous* (that is, time can be represented either by natural or rational or real numbers, but not by numbers whose cardinality is higher than the reals).

  2. $<$ the earlier/later relation. It is irreflexive, anti-simmetric, transitive, backward linear and forward branching. Maximal linearly ordered subsets of T are called *Histories* (h, h', h'', ...). $H_t := \{h \mid t \in h\}$.

- J is a partition of T. The elements i, i', ... of J are are pairwise disjoint sets of moments and are called *instants*. $i(t)$ is the instant containing the moment t.

- Ag is a finite set of agents. In our case, $Ag := \{a\}$.

- *Choice* is a function assigning to pairs $\langle t, a\rangle$ a partition $Choice_{\langle t,a\rangle}$ of H_t. The elements of such a partition are denoted by $[h]^{Choice}_{\langle t,a\rangle}$, $[h']^{Choice}_{\langle t,a\rangle}$ and are called *possible choices of a at t*.

Histories satisfy the following:

H1 historical connection: Any two histories overlap at some moment. Two histories h and h' overlap at t if $t \in h \cap h'$.

H2 isomorphic temporal order: All histories in a tree share an isomorphic temporal ordering.

The moments that are in $i(t)$ are to be thought as "as the set of alternate possibilities for 'filling' the time of t" (Belnap, 1991, p. 144). Saying it with other words, they are "contemporaneous [with t] in different histories", i.e., they should be assumed to take "the same time" as t to realise (in different course of events). As Belnap writes: "I need instants because I think that for the sense of stit I am after, in considering whether Autumn Jane stit she was muddy at a certain moment, it is relevant to consider what else might have been at the instant inhabited by that moment" (Belnap, 1991, p. 144). We shall come back to their nature below. Instants satisfy:

J1 a history, a moment: There is only a moment t such that $t \in h \cap i$ (h and i arbitrary).

J2 order preservation: Given any two histories h and h', the set of ordered pairs $\{\langle h \cap i, h' \cap i\rangle \mid i \in J\}$ is an order preserving bijection from h onto h'.

The *possible choices* of a at t represent the classes of the developments of the events that is in a's power at t to undertake. Each possible choice satisfies these conditions:

Ch1 no choice without dividedness: If h and h' are undivided at t, then $[h]^{Choice}_{\langle t,a\rangle} = [h']^{Choice}_{\langle t,a\rangle}$.

where two histories h and h' are undivided at t iff there is some $t' > t$ such that $t' \in h \cap h'$.

To evaluate the formulae in **AS** we use a stit-model $\mathcal{M}^{stit} := \{\mathcal{T}^{stit}, V\}$, where V in an evaluation function from formulae to pairs moments/histories. The truth-clauses for the formulae in **AS** are as follows:

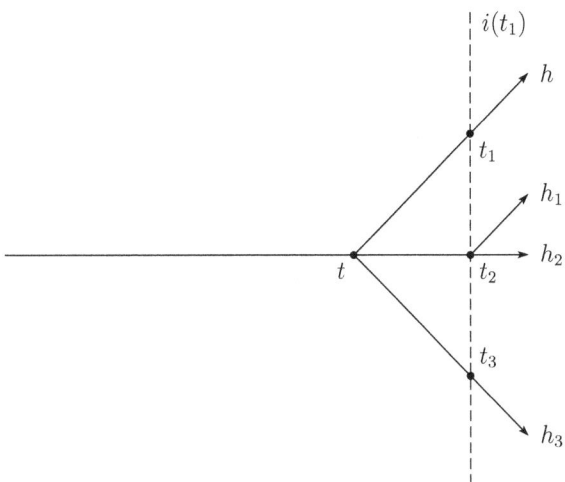

Figure 1: A stit-tree $[h]^{Choice}_{\langle t,a \rangle} = \{h, h_1, h_2\}$, $[h']^{Choice}_{\langle t,a \rangle} = \{h_3\}$.

AS1 $\mathcal{M}^{stit}, \langle t, h \rangle \vDash p$ iff $\langle t, h \rangle \in V(p)$.

AS2 $\mathcal{M}^{stit}, \langle t, h \rangle \vDash astit_a\, A$ iff (PR) there is a $t' < t$ and such that A is true at all those moments t'' that are both in $i(t)$ and belong to a history h' that is in $[h]^{Choice}_{\langle t',a \rangle}$, and (NR) A is false at least a moment t''' that belongs to $i(t)$ and is in a history h'' that is *out* of $[h]^{Choice}_{\langle t',a \rangle}$.

When a moment t'' is in $i(t)$ and belongs to a history $h' \in [h]^{Choice}_{\langle t',a \rangle}$ for some t', I shall say that t'' is "choice-equivalent to t at t'". A moment t''' that is in $i(t)$ and does not belong to any history $h' \in [h]^{Choice}_{\langle t',a \rangle}$ for some t', I shall call "a counter of t at t'''". The clauses for the boolean constructions on atoms are straightforward. *Stit-frames* are defined out of *stit-models* in the standard way. Truth in a stit-model ($\mathcal{M}^{stit} \vDash A$), in every stit-model (in a stit-tree, $\mathcal{T}^{stit} \vDash A$) and validity (**AS** $\vDash A$) are defined as usual. Axioms for a mono-agent logic with achievement stit are given in (Xu, 1994) and in (Belnap et al., 2001, pp. 383–384). Here, we will not need any reference to a specific axiom system for **AS**. However, since any logic based on stit languages

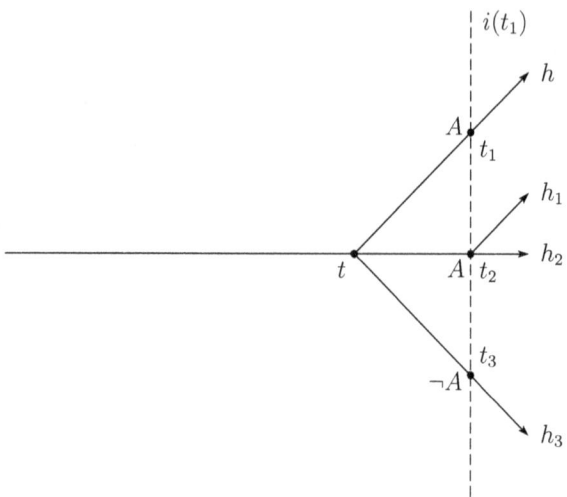

Figure 2: $astit_a\ A$ is true at $\langle t_1, h\rangle$, $[h]^{Choice}_{\langle t,a\rangle} = \{h, h_1, h_2\}$, $[h']^{Choice}_{\langle t,a\rangle} = \{h_3\}$.

is conceived as an extension of classical logic, here we assume that any classical axiom and rules of inference hold in **AS**.

In **AS1**, PR ("the positive requirement") implies that, in order for $astit_a\ A$ to be true at $\langle t, h\rangle$ (in the intended model), there must be a witness (t') earlier than t (the latter being the moment of evaluation) such that every moment t'' that is choice-equivalent to t at t' is such that $\mathcal{M}^{stit}, \langle t'', h'\rangle \models A$ for every history h' passing through t''. The intuitive meaning of PR is clear: in order for a to see to it that A at $\langle t, h\rangle$, at a previous moment a must determine the events to lie into courses where A is true (where the relevant courses are in $[h]^{Choice}_{\langle t',a\rangle}$). By "determining A to hold" here I mean that, at a given pair $\langle t, h\rangle$, A is true at every $h' \in [h]^{Choice}_{\langle t',h'\rangle}$ and some moment $t' \in h'$ and such that $t < t'$. PR fits our intuitions: if a did not manage to determine the events in that way, we would not say that a sees to it that A. NR ("the negative requirement") implies that there is a history h'' where A fails to happen at a counter of t at t'. The reason for NR is evident: if A would have happened in any case (hence without a possibility of failure), we would not say that the holding of A is due to the choice and consequent action of a. A consequence of NR is that the theoremhood og A implies the falsehood of $astit_a\, A$ does not hold: if A is a theorem, it cannot satisfy NR, and hence $astit_a\, A$ cannot be true. As a consequence, **AS** is a non-normal logic, that is a modal logic where the rule of inference $A/\, astit_a\, A$ does not hold. This fits with the idea that it makes no sense to say that we see to it that logical laws hold.

As I said in the introduction, $(astit_a(A\to B)\wedge astit_a\, A)\to astit_a\, B$ is not valid. Suppose both $astit_a\, A\to B$ and $astit_a\, A$ are true at $\langle t,h\rangle$; as a witness, the former has t', while the latter has t'', with $t' < t''$. Assume $\neg B$ is false at t. Hence t' fails to satisfy PR for $astit_a\, B$. Assume also that B is true for every history h' passing through t'' and any moment that is in $i(t)\cap h'$. Hence t'' fails to satisfy NR for $astit_a\, B$. As a consequence, $astit_a\, B$ is false at $\langle t,h\rangle$. The proof slightly modifies the one given in (Horty & Belnap, 1995, p. 598) to show that *astit* is not closed under Modus Ponens. For reasons of space, I put no figure for representing the counterexample to axiom K. A figure to this purpose can be found in (Horty & Belnap, 1995, p. 598). The counterexample relies on the fact that *different* witnesses may be relevant for the evaluation of different agentives. Indeed, **AS2** does not imply that *one given moment* should be the witness of one or many agentives.

3   Metric Possible Choices

The achievement stit operator I propose is to be defined in the tempo-modal language **MPC**. Such a language differs from the standard languages from branching time since (1) it is a metric language;[3] (2) some of its operators express a quantification over possible choices, and one of them expresses a restricted quantification, which ranges over the histories of an intended possible choice. The first step to **MPC** and n-$astit_a$ is the definition of the duration function dur that allows **MPC** to express the idea of a distance between moments.[4] The function as-

---

[3] For an introduction to metric tempo-modal logics, see (Øhrstøm & Hasle, 1995, pp. 231–235 and 382).

[4] A duration function is not the only way to guarantee the ability of expressing metric notions. For other ways, see (Øhrstøm & Hasle, 1995). A duration relation is introduced in (Øhrstøm, 2009) to ground the semantics of the metric language presented there. However, the conditions I shall give below are not presented in (Øhrstøm, 2009).

signs a number to each pair $\langle t, t' \rangle$ of *comparable different* moments. That number represents the distance between the two moments.

Definition 1 (Def. *dur*). *dur* is a function $T \times T \longmapsto \mathbf{X} - 0$ (where $\mathbf{X}$ is either the set of the positive naturals, or of the positive rationals, or of the positive reals), and such that:

1. **no *dur* between t and t**. This is to avoid 0 as a value of the function, or to avoid $dur(t,t) = n > 0$.

2. **symmetry of distance**: $dur(t,t') = dur(t',t)$. This is to avoid that the distance between t and t' is assigned another value than the distance between t' and t.

3. **Given any h, just a t' is such that** $dur(t,t') = x$ and $t < t'$ ($t' < t$). This is to guarantee that, in a given history, just a moment t' is later (earlier) than t and has value x for its distance from t.

Given any arbitrary moment $t \in h$, I say that the only moment $t' \in h$ such that $dur(t,t') = x$ and $t < t'$ is "n-later than t". Analogously, given any arbitrary moment $t \in h$, I say that the only moment $t' \in h$ such that $dur(t,t') = x$ and $t' < t$ is "n-earlier than t".

We can now introduce the language of Metric Possible Choices. **MPC** is the language of classical logic **PL** plus the operator $\Diamond^C$ and infinite operators of the form $\mathbf{F_n}$, $\mathbf{f_n}$, $\mathbf{P_n}$. Their duals are $\Box^C$ and infinite operators of the form $\mathbf{g_n}$, $\mathbf{G_n}$, $\mathbf{H_n}$, respectively. The index $\mathbf{n}$ is a *meta-term* for numbers, which in turn express the value of *dur* between the moment of evaluation and the moment that is (moments that are) relevant for the evaluation. This means that $\mathbf{n}$ may stay for 1, 2, or any value admitted by *dur*. In a given context, $\mathbf{n}$ does not vary its meaning and hence in a formula such as $\mathbf{F_n}A \wedge \mathbf{P_n}A$ both occurrences of n stay for the same number.[5] It is obvious that this would not hold if $\mathbf{n}$ were a variable. The intuitive meaning of the operators is as follows: $\mathbf{F_n}A$ ($\mathbf{f_n}A$) expresses the idea that for every (some) history h' in the intended possible choice of a at t, A is true at $\langle t', h' \rangle$, where $dur(t,t') = n$, $t < t'$ and $t' \in h'$. $\mathbf{P_n}A$ expresses the idea that in the intended possible choice of a at t, A is true at a moment

---

[5] Just to give an idea, $\mathbf{n}$ and the other meta-terms for numbers behave like the meta-terms A and B in the expression of the inference rule $A \wedge B / A$.

astit and MPC

$\langle t', h \rangle$, where $dur(t, t') = n$ and $t' < t$. $\Diamond^C A$ expresses that there is a possible choice $[h']^{Choice}_{\langle t,a \rangle}$ such that A is true at the intended moment t with respect to $[h']^{Choice}_{\langle t,a \rangle}$. $\Diamond^C A$ is the only non-metric operator of **MPC**, and quantifies over *possible choices*. $\mathbf{F_n}$, $\mathbf{f_n}$, $\mathbf{P_n}$ contain both a metric and a quantification over histories in a given possible choice. Both kinds of operators constitute a significant technical novelty with respect to the classical languages for branching time, whose modal operators unrestrictedly quantify over histories, simply.

The formulae in **MPC** are evaluated on the basis of metric-trees of the form $\mathcal{T}_{\mathbf{MPC}} := \langle \mathcal{T}, dur, Ag, Choice \rangle$, where $\mathcal{T}$, Ag and $Choice$ are as in $\mathcal{T}^{stit}$, and dur is defined as above. To evaluate the formulae in **MPC** we use a metric-model $\mathcal{M}_{\mathbf{MPC}} := \{\mathcal{T}_{\mathbf{MPC}}, \mathcal{V}_{\mathbf{MPC}}\}$, where V in an evaluation function from formulae to pairs moments/histories. The truth-clauses for the tempo-modal sentences of **MPC** are:

MPC1 $\mathcal{M}_{\mathbf{MPC}}, \langle t, h \rangle \models \mathbf{F_n} A$ iff for all $h' \in [h]^{Choice}_{\langle t,h \rangle}$, there is a t' such that $t < t'$ and $t' \in h$ and $\mathcal{M}_{\mathbf{MPC}}, \langle t', h \rangle \models A$ and $dur(t, t') = n$

MPC2 $\mathcal{M}_{\mathbf{MPC}}, \langle t, h \rangle \models \mathbf{f_n} A$ iff there are an $h' \in [h]^{Choice}_{\langle t,h \rangle}$ and a t' such that $t < t'$ and $t' \in h$ and $\mathcal{M}_{\mathbf{MPC}}, \langle t', h \rangle \models A$ and $dur(t, t') = n$

MPC3 $\mathcal{M}_{\mathbf{MPC}}, \langle t, h \rangle \models \mathbf{P_n} A$ iff there is a $t' < t$ and such that $\mathcal{M}_{\mathbf{MPC}}, \langle t', h \rangle \models A$ and $dur(t, t') = n)$

MPC4 $\mathcal{M}_{\mathbf{MPC}}, \langle t, h \rangle \models \Diamond^C A$ iff $\exists [h']^{Choice}_{\langle t,a \rangle}$ such that $\mathcal{M}_{\mathbf{MPC}}, \langle t, h' \rangle \models A$

The clauses for sentences containing no tempo-modal operators are straightforward. Truth in a metric-model ($\mathcal{M}_{\mathbf{MPC}} \models A$), every metric-model (in a metric-tree, $\mathcal{T}_{\mathbf{MPC}} \models A$) and **MPC**-validity (**MPC** $\models A$) is defined as usual. Notice that, by dur, $\mathbf{F_n} A \leftrightarrow \mathbf{G_n} A$, $\mathbf{f_n} A \leftrightarrow \mathbf{g_n} A$ and $\mathbf{P_n} A \leftrightarrow \mathbf{H_n} A$ are **MPC**-valid. Here, I do not give any axiom system to **MPC**, but I assume that all classical axiom and rules of inference hold in **MPC**.

It is now time to define the operator **n**-$astit_a$:

Definition 2 (n-*astit*). n-$astit_a\, A :=_{\mathbf{MPC}} \mathbf{P_n}(\mathbf{F_n} A \wedge \Diamond^C \mathbf{f_n} \neg A)$

The right side of the definition reads: at the duration **n** in the past ($\mathbf{P_n}$) it was true that: (i) A is true at a duration **n** in the future

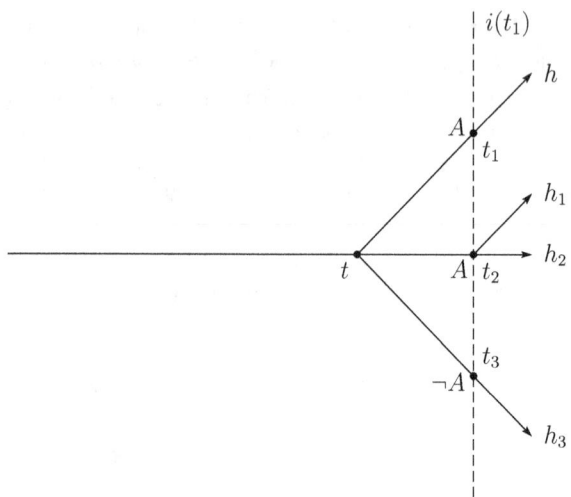

Figure 3: $\mathbf{n}$-$astit_a\, A$ is true at $\langle t_1, h\rangle$, $[h]_{\langle t,a\rangle}^{Choice} = \{h, h_1, h_2\}$, $[h']_{\langle t,a\rangle}^{Choice} = \{h_3\}$, $dur = \mathbf{n}$

w.r.t. every history of the intended possible choice ($\mathbf{F_n}A$), and (ii) A is false at a duration $\mathbf{n}$ in the future in a history of some possible choice passing through the intended moment t ($\Diamond^C \mathbf{f_n} \neg A$). It is easy to see that, in **MPC**, it holds that $\mathbf{P_n}(\mathbf{F_n}A \wedge \Diamond^C \mathbf{f_n} \neg A) \leftrightarrow \mathbf{P_n F_n}A \wedge \mathbf{P_n}\Diamond^C \mathbf{f_n}\neg A$. This helps us seeing that $\mathbf{P_n F_n}A$ corresponds to PR, and $\mathbf{P_n}\Diamond^C \mathbf{f_n}\neg A$ corresponds to NR.

Though apparently complex, the intuitive meaning of $\mathbf{n}$-$astit_a$ is simple. Suppose the moments in our **MPC**-models are days, and $\mathbf{n}$ stays for "5". Then "$\mathbf{n}$-$astit_a\, A$" means "5 days ago a determined A to hold, and A was susceptible of failing". Here, the locution "a determines A to hold (in $\mathbf{n}$ time units)" is the informal expression of the **MPC** formula $\mathbf{F_n}A$, while the locution "A is susceptible of failing (in $\mathbf{n}$ moments)" is the informal expression of the **MPC** formula $\Diamond^C \mathbf{f_n}\neg A$.

The reason for the introduction of $\mathbf{n}$-$astit_a$ is that the operator $\mathbf{n}$-$astit_a$ makes Axiom K valid. An easy proof can be given of

$$\mathbf{MPC} \vDash (\mathbf{n} - astit_a(A \to B) \wedge \mathbf{n} - astit_a A) \to \mathbf{n} - astit_a B.$$

Proof. Suppose $\mathbf{n}\text{-}astit_a(A \to B) \land \mathbf{n}\text{-}astit_a\, A$ is true at an arbitrary pair $\langle t, h\rangle$. The truth of $\mathbf{MPC} \models \mathbf{n}\text{-}astit_a(A \to B)$ at $\langle t, h\rangle$ implies that (A) there is a moment t' $\mathbf{n}$-earlier than t; (B) in any arbitrary history $h' \in [h]^{Choice}_{\langle t', a\rangle}$, the moment t'' $\mathbf{n}$-later than t' is such that $A \to B$ is true at $\langle t'', h'\rangle$; (C) there is a history h'' where the moment t''' $\mathbf{n}$-later than t' is such that $A \land \neg B$ is true at $\langle t''', h''\rangle$. The truth of $\mathbf{MPC} \models \mathbf{n}\text{-}astit_a\, A$ at $\langle t, h\rangle$ implies (A) as above and (D) in any arbitrary history $h' \in [h]^{Choice}_{\langle t', a\rangle}$ the moment t'' $\mathbf{n}$-later than t' is such that A is true at $\langle t'', h'\rangle$; for any $h' \in [h]^{Choice}_{\langle t', a\rangle}$, such a moment is the same as the moment that is mentioned in (B): there is indeed only one moment such that it belongs to h' and is $\mathbf{n}$-later than t' (see *dur*); (E) there is a history h'' where the moment t''' $\mathbf{n}$-later than t' is such that $\neg A$ is true at $\langle t''', h''\rangle$. Now suppose that $\mathbf{n}\text{-}astit_a\, B$ is false at $\langle t, h\rangle$. By **Def.** $\mathbf{n}\text{-}astit$, (C) would guarantee that the falsehood of the formula would not be due to $\Box^C \mathbf{G_n} B$ (at $\langle t', h\rangle$), since B is false at $\langle t''', h''\rangle$. Thus, the falsehood of the formula should be due to $\mathbf{f_n} \neg B$ (at $\langle t, h\rangle$). This implies that (F) there is a history $h' \in [h]^{Choice}_{\langle t', a\rangle}$ where the moment t'' $\mathbf{n}$-later than t' is such that B is false at $\langle t'', h'\rangle$. For any arbitrary $h' \in [h]^{Choice}_{\langle t', a\rangle}$ such a moment is the same as the moment that is mentioned in (B), since there is only one moment such that it belongs to h' and is $\mathbf{n}$-later than t'. By (B), (D) and (F) we should have $A \to B$, A and $\neg B$ at $\langle t'', h'\rangle$, thus violating Modus Ponens. But this contradicts our assumption that classical rules of inference are valid in $\mathbf{MPC}$.

4 Concluding Remarks

In the present paper, I have proposed a new achievement stit operator, $\mathbf{n}\text{-}astit_a$ (being a metric achievement operator), and I have proved the technical advantage connected by it: the new operator makes axiom K valid, while the usual stit operator does not. Before concluding, I would like to stress two conceptual virtues of the proposal: the ability of distinguishing determination from inevitability and the ability of reducing agentives to tempo-modal formulae. These issues in turn highlight some distinctive connections between stit-sentences and branching-time, as well as some distinctive features of the philosophy of agency of Belnap et al.

Determination without inevitability. A central standpoint of the account of agency of Belnap et al. is that there is a difference between *determination* (as characterised in sections 2 and 3) and *inevitablity*. The first notion refers to the holding of something that is due to the intervention of the agent. In our branching-time framework, this corresponds to a given formula being true in any history h in the intended possible choice and in any moment in h. The second notion refers to the holding of something no matter how agents do or other events go. In our branching-time framework, this corresponds to a given formula being true in every history h at some moment. In standard tempo-modal languages like the ones introduced in (Prior, 1967) and (Thomason, 1984), there is no chance to express such a difference. The modal operators of such languages express an unrestricted quantification over histories, and as a consequence they cannot single out a given set of histories (the histories in a given possible choice, say) when evaluating a formula. In the language of Metric Possible Choices, the difference may be expressed: *determining* A is expressed by $\mathbf{F_n}A$ and *inevitably* A is expressed $\square^C \mathbf{F_n}A$. If we allow quantifiers over numbers, we get an expansion of **MPC** that is able to express the above difference without reference to a specific moment. In this case, *determining* A is expressed by $\exists x \mathbf{F}_x A$ and *inevitably* A is expressed $\square^C \exists x \mathbf{F}_x A$. The importance of such a difference is due to (NR), which stresses that, in order for A to be done by an agent, A must be determined, without being inevitable (see section 2 and 3). As a consequence, the difference I have drawn could be unappreciated within those approaches to stit that does not take (NR) as an indispensable standpoint of the philosophy of agency (for example, the approach by Chellas, see (Belnap et al., 2001) and (Chellas, 1992)).

A logic for agentives without agentives? My approach brings at the linguistic level what was confined at the ontological level in the traditional approach to achievement stit: the intrinsic temporal character of the achievement stit operator. The paper has presented a metric achievement stit operator, *defined via metric tempo-modal operators*. As a consequence, **n**-$astit_a$ A is nothing but an abbreviation for a tempo-modal formula. This makes agentives dispensable. Thus, a question naturally rises: is **AS** -so to speak- a logic for agentives without agentives? All in all, agentives can be reduced to formulae that do not contain any agency concept. The present proposal has given, in the author's opinion, a significant hint on the possibility of reducing some logics of the Stit framework (at least **AS**) to a tempo-modal logic. The extension of **MPC** to a logic with many agents is easy and it is likely that it would express a multi-agent version of the logic **AS**. If this is true, an important logic of agency can be reduce to a refined tempo-modal logic. Yet, in the author's opinion, this does not unveil a conceptual inadequacy of **AS**. It just tells us what already emerges by the approach of Belnap et al.: **AS** and all the other Stit Logic do not (and in fact cannot) give us an *analysis* of action and agentives. However, this does not prevent **AS** (and the Stit framework in general) from giving us a suitable and successful *representation-tool* for modelling the interaction between actions, available choices, and time. And this does not stop Stit logic from playing the prominent role it is playing in the logic of agency since the last fifteen years.

Roberto Ciuni
Section of Philosophy, Delft University of Technology
Jaffalaan 5 2628 BX Delft
r.ciuni@tudelft.nl, ciuniroberto@yahoo.it

References

Belnap, N. (1991). Before refraining: Concepts for agency. *Erkenntnis*, *34*, 137–169.

Belnap, N., & Perloff, M. (1992). The way of the agent. *Studia Logica*, *51*, 463–484.

Belnap, N., Perloff, M., & Xu, M. (2001). *Facing the Future: agents and choices in our Indeterminist World*. Oxford: Oxford University Press.

Chellas, B. (1992). Time and modality in the logic of agency. *Studia Logica*, *51*, 485–517.

Horty, J., & Belnap, N. (1995). The deliberative stit: A study of action, omission, and obligation. *Journal of Philosophical Logic*, *24*(6), 583–644.

Øhrstøm, P. (2009). In defence of the thin red line: A case for ockhamism. *Humana.mente*, *8*, 17–32.

Øhrstøm, P., & Hasle, P. (1995). *Temporal logic: From ancient ideas to artificial intelligence.* Dordrecht: Kluwer.

Prior, A. (1967). *Past, present and future.* Oxford: Oxford University Press.

Thomason, R. (1984). Combination of tense and modality. In D. Gabbay & F. Guenthner (Eds.), *The handbook of philosophical logic* (pp. 135–165). Dordrecht: Reidel.

Xu, M. (1994). Decidability ofstit theory with a single agent and refref equivalence. *Studia Logica, 53*(2), 259–298.

Zanardo, A. (1998). Undivided and indistinguishable histories in branching-time logics. *Journal of Logic, Language and Information, 7,* 297–315.

Zanardo, A. (2009). Modalities in temporal logic. *Humana.mente, 8,* 1–15.

Surprises in Logic

Catarina Dutilh Novaes*

1 Introduction

In the *Tractatus* (6.1251), Wittgenstein (in?)famously said: "Hence, there can never be surprises in logic." One way of understanding this claim is to view Wittgenstein as embracing one of the horns of what is often referred to as the 'paradox of inference' (which is in fact a dilemma rather than a paradox properly speaking): the tension between the validity and the usefulness of logic, and of deductive reasoning more generally.[1]

> If in an inference the conclusion is not contained in the premises, it cannot be valid; and if the conclusion is not different from the premises, it is useless, but the conclusion cannot be contained in the premises and also possess novelty; hence [deductive] inferences cannot be both valid and useful. (Cohen & Nagel, 1934, p. 173)

In the *Tractatus*, Wittgenstein clearly chooses validity over usefulness in his account of logic. As portrayed by him, logic is essentially uninformative: logical propositions "say nothing" (6.11), as they do not depict (contingent) states-of-affairs; "we can actually do without logical propositions" (6.122).[2] Indeed, his conclusion to the effect that there cannot be surprises in logic follows neatly from the premises he lays down in section 6.1 of the *Tractatus*. So whoever wishes to find a different solution to the dilemma, and in particular to give an

---

*Thanks to Martin Stokhof for comments on an earlier draft. I must also warn the reader that the ideas presented here are very much work in progress; many of the claims made here shall not receive a full treatment for now.

[1] Notice that the name 'paradox of inference' is also often used in connection with Lewis Carroll's puzzle of Achilles and the Tortoise. This usage of the expression seems to stem from (Clark, 2002), and it concerns a different issue, namely the threat of infinite regress for a rule-following conception of (logical) reasoning.

[2] But in all fairness, Wittgenstein's position is more subtle than these remarks may seem to suggest. In particular, even though logical propositions are 'senseless', logic is a *Spiegelbild* of the world (6.13).

account of logic which allows for validity and usefulness to co-exist, must address Wittgenstein's argument in order to block the conclusion: either at least one of the premises does not hold, or at least one of the inferential steps is not legitimate (so naturally, there are several ways of blocking Wittgenstein's argument). Here I shall focus on one particular aspect of the argument, namely his characterization of the epistemology of what he calls a 'mechanical expedient', the derivation of proofs. I agree with Wittgenstein's claim that what could be described as a mechanical way of reasoning — calculation — is a significant aspect of logical practice (though I certainly do not wish to imply that it is the only or not even the most important aspect thereof); but I question his characterization of this way of reasoning as delivering no surprises.

For this purpose, I draw on some experimental results from the psychology of reasoning, which suggest that a very pervasive mechanism in our ordinary, everyday life ways of reasoning is a tendency to seek confirmation of previously held beliefs; this mechanism is referred to as 'confirmation bias' in the psychology literature. In particular, it is identified as the source of the empirically observed phenomenon of people tending to endorse arguments with plausible conclusions (i.e. conclusions according with their prior beliefs) far more frequently than arguments with implausible conclusions, regardless of the argument's validity as such. In the case of argument endorsement, the term typically used is 'belief bias', which is a special case of confirmation bias. Indeed, in experiments *in*valid arguments with plausible conclusions are more often endorsed than valid arguments with *im*plausible conclusions. My suggestion here will be that the 'mechanical' manner of reasoning often used in logic (and other deductive fields) in fact acts as a counterbalance to our tendency towards confirmation and belief bias, and that this is one of the reasons why deductive reasoning can be and often is informative after all. The built-in belief bias suppressing effect of deductive reasoning (which is particularly acute in the case of deductive 'mechanical' reasoning) makes it in fact particularly apt to deliver surprises.

I shall first discuss the so-called 'paradox of inference' and some of the different treatments it received in the literature. Next, I turn to the *Tractatus* in order to spell out the structure of the argument that led Wittgenstein to conclude that there can be no surprises in logic. I then discuss the mechanism of belief bias as described in the psychology of reasoning literature. Finally, I argue that 'mechanical' reasoning can serve as a counterbalance to belief bias and thus that its use in logic can partially explain why there seem to be surprises in logic after all.

2 The paradox of inference and surprises in logic

The so-called paradox of inference has received quite some attention from logicians and philosophers at different times, and rightly so: it is perhaps the most crucial philosophical issue concerning deductive reasoning in general. The point of deductive reasoning is precisely not to allow for external information to 'sneak in', to conduct reasoning solely on the basis of the information explicitly on the table in such a way that no additional information could possibly defeat the conclusion at a later stage. But if all the information that can be used and is relevant is already on the table, how can we learn anything new by reasoning deductively? In a sense, the conclusion is already contained in the previously available information, the premises. "The existence of deductive inference is problematic because of the tension between what seems necessary to account for its legitimacy and what seems necessary to account for its usefulness." (Dummett, 1978, p. 297) (Notice that this discussion intersects directly with the matter of analyticity, which is of course one of the concepts used to explain in more precise terms what it means for the conclusion to be 'contained' in the premises — see (Primiero, 2008).)

Awareness of this somewhat paradoxical nature of deductive reasoning can be traced back to Mill (Mill, 1843). Some of the authors who have taken an interest in the issue are Keynes (Keynes, 1884, p. 414 et passim), Hintikka (Hintikka, 1973, p. 222 et passim), and more recently D'Agostino and Floridi (D'Agostino & Floridi, 2009), Sequoiah-Grayson (Sequoiah-Grayson, 2008) (the last three refer to the issue as 'a scandal of deduction', a term introduced by Hintikka) and Primiero (Primiero, 2008, sec. 2.3), among others. Different approaches and solutions to the dilemma have been proposed in order to explain why there seems to be gain of information when one performs a deduction, even though the whole point is precisely not to go beyond what is given by the premises. Hintikka proposed a syntactic approach to the issue, based on the notion of distributive normal form of a sentence F; Sequoiah-Grayson criticized Hintikka's solution and suggested that instead a semantic approach would be required to deal with the puzzle. D'Agostino and Floridi address the issue from the point of view of the computational complexity of different logical systems. They notice that there is a certain tension between the idea that logic does not yield new information and the undecidability of first-order logic, and suggest that, on the basis of some computational properties, a hierarchy of logical systems can be formulated, which would determine the lower or higher degree of 'analyticity' of a given system. Primiero addresses the issue from a constructive point of view, arguing that the tension between the validity and the usefulness of logic can be dissipated once one pays sufficient attention to the distinction between proposition and judgment.

I think these different solutions all outline important factors that may be involved in the apparent information gain resulting from reasoning logically (deductively), and outlining yet another one of such factors is the purpose of the present contribution. In other words, I believe that there are different phenomena involved, and thus that different accounts may complement each other rather than compete with each other when it comes to explaining why there seems to be information gain in logic after all. To my knowledge, the 'paradox of inference' has never been formulated specifically in terms of the presence or absence of surprises in logic, but some authors (Carnap, Hintikka) come very close to this idea when pointing out that information flow and information gain go hand in hand with unpredictability (e.g. that the measure of content of a proposition is inversely proportional to the probability weight assigned to it). There is information gain in particular when the outcome of some reasoning goes against what was initially anticipated — thus, when the conclusion is unexpected and surprising in one way or another. Indeed, the paradox (dilemma) can also be neatly formulated in terms of surprises: for deductive reasoning to be legitimate, it must not deliver any surprises; but for deductive reasoning to be useful, it must do just that — deliver surprises.[3]

---

[3] Notice however that 'surprise' is a stronger notion than 'information gain': there may be information gain even if the new information is not particularly surprising, i.e. if the agent involved did not have specific expectations concerning it prior to obtaining it.

3 Wittgenstein on surprises and 'mechanical reasoning'

Wittgenstein's claim that there can be no surprises in logic is of course related to a deeper feature of the philosophy of logic defended in the *Tractatus*: his realist vision of logic, as opposed to what could be described as a Fregean, epistemological account of logic as a tool to acquire new knowledge. The search for an ontological grounding for logic (and in fact for language in general) leads Wittgenstein away from matters such as the potential cognitive role of logical reasoning. So in a sense, it is to be expected that he would be prepared to 'bite the bullet' of logic's uselessness. Still, it is interesting to take a closer look at the argument leading to the conclusion that there can be no surprises in logic in order to see what exactly the conclusion depends on.

The cornerstone of the Tractarian philosophy of logic is stated at the very beginning of the section we will be concerned with here, namely section 6.1:

> 6.1 The propositions of logic are tautologies.

On the one hand, to limit the scope of logic in this way may seem unwarranted, as logic presumably also deals with hypothetical reasoning starting with open assumptions. On the other hand, this may also simply be a different formulation to the familiar slogan according to which logic (and deductive reasoning more generally) is 'tautological' precisely in the sense spelled out above, i.e. in the sense that the conclusion of a logical argument is already contained in the premises. Moreover, by the deduction theorem (which holds in most, but not all, logical systems), any valid argument can be formulated as a conditional having as antecedent its premises and as consequent its conclusion, and such a conditional is indeed a tautology in the sense of 'true no matter what'. (Notice also that proofs in sequent calculus are trees whose initial nodes are all instances of the identity axiom $A \to A$.) So there is nothing particularly controversial in Wittgenstein's claim, even though one could probably offer an account of logic that is not limited to tautological propositions.

That the propositions of logic are all tautological does not yet imply that there shall be no surprises in logic; to be a tautology is a *factual* (albeit necessary) property of propositions, whereas the phenomenon of surprises concerns the *recognition* by an agent that some object a has property P (in this case, that a proposition is a tautology). The feature of logical propositions that really excludes the possibility of surprises is a different one: for Wittgenstein, every tautological proposition *shows* its own 'tautologyness'. Just as he insists that the truth or falsity of non-logical propositions is a purely contingent matter (determined by the existence or non-existence of the situation that is its sense), a key feature of logical propositions is that their truth (or falsity, in the case of contradictions) can be 'read off' from the propositions directly.

> 6.113 It is the peculiar mark of logical propositions that one can recognize that they are true from the symbol alone, and this fact contains in itself the whole philosophy of logic. And so too it is a very important fact that the truth or falsity of non-logical propositions cannot be recognized from the propositions alone.

> 6.127 All the propositions of logic are of equal status: it is not the case that some of them are essentially derived propositions. Every tautology itself shows that it is a tautology.

How can a proposition show itself to be true (and necessarily so)? To understand this claim, one must bear in mind that, for Wittgenstein in the *Tractatus*, validity in logic is purely determined by logical syntax, i.e. by the (internal) properties and possibilities of the signs themselves, viewed as objects as such, thus not in their representative dimension (e.g., as pictures of something else, precisely because logical propositions do not depict possibilities). So in a sense, the objects that a logical proposition 'speaks of' are exclusively signs, in particular those very signs being presented, and this is why its truth can be determined by mere inspection of the proposition. Thus, it would appear that, for Wittgenstein (of the *Tractatus*) just as for Peirce, logic is the science of (the necessary laws of) signs.

> 6.124 [...] We have said that some things are arbitrary in the symbols that we use and that some things are not. In logic it is only the latter that express: but that means that logic is not a field in which we express what we wish with the help of signs, but rather one in which the nature of the absolutely necessary signs speaks for itself. If we know the

logical syntax of any sign-language, then we have already been given all the propositions of logic.

But what about proof and derivation? Isn't the practice of formulating proofs essential for the establishment of the (necessary) truth of a logical proposition? Or does Wittgenstein deny that formulating proofs has always been part and parcel of logic as an enterprise? Well, he does have an account of proofs in logic, but one which significantly downplays their importance:

> 6.126 One can calculate whether a proposition belongs to logic, by calculating the logical properties of the symbol. And this is what we do when we 'prove' a logical proposition. For, without bothering about sense or meaning, we construct the logical proposition out of others using only rules that deal with signs. The proof of logical propositions consists in the following process: we produce them out of other logical propositions by successively applying certain operations that always generate further tautologies out of the initial ones. (And in fact only tautologies follow from a tautology.) Of course this way of showing that the propositions of logic are tautologies is not at all essential to logic, if only because the propositions from which the proof starts must show without any proof that they are tautologies.
>
> 6.1262 Proof in logic is merely a mechanical expedient to facilitate the recognition of tautologies in complicated cases.

Wittgenstein attributes a 'mechanical' character to the operations of transforming propositions into others — i.e. of inferring A from B. Importantly (for my purposes), he says that when conducting a proof one proceeds "without bothering about sense or meaning, [...] using only rules that deal with signs". This observation fits in well with the idea that logic is the science of signs as such, i.e., not insofar as they are pictures of something else outside them (which would be their potentially meaningful dimension). But if every tautology shows indeed itself to be a tautology, proof is a rather superfluous device, of purely instrumental value. This again confirms Wittgenstein's realist perspective: logic is concerned with the objective properties of signs, which are all there from the start (just as the internal properties of objects as described in the first section of the *Tractatus*). The 'merely' epistemological layer of an agent actually uncovering (unfolding) these properties, which is what happens when she formulates a proof, is unimportant. The properties are already (or should be, in any case) shown in a logical proposition, which shows itself to be a tautology.

> 6.1265 It is always possible to construe logic in such a way that every proposition is its own proof.

But because the objective properties of signs (or their 'necessary laws', as Peirce would say) are a given, i.e. are fully ontologically determined, the class of logical propositions is also fully determined from the start, much before any proof is formulated.

> 6.125 It is possible — indeed possible even according to the old conception of logic — to give in advance a description of all 'true' logical propositions.

And this is indeed the aphorism immediately preceding the claim under scrutiny here:

> 6.1251 Hence, there can never be surprises in logic.

Thus, from the start, the whole of logic is given to us, since the objective properties of the signs are determined from the outset. It is all there 'on the table', as it were, and if we fail to see it immediately, this can only be due to our own inability or perhaps to a defective notation (one which fails to formulate every tautology as showing its own tautological nature). Wittgenstein's ontic point of view entails a disregard for the actual *process* of 'unfolding' propositions and coming to recognize them as tautologies by means of a proof, given that the *result* of a proof, i.e. the conclusion of the derivation, is in itself already a 'proof' of its own truth.

> 6.1261 In logic process and result are equivalent. (Hence the absence of surprise.)

There are of course many aspects of Wittgenstein's characterization of logic one could object to in order to block the argument leading to the conclusion that there can be no surprises in logic. I shall focus on contesting the following claim: "6.1262 Proof in logic is merely a mechanical expedient to facilitate the recognition of tautologies in complicated cases." The idea seems to be that what is achieved by means of a proof could be achieved without a proof,[4] and Wittgenstein seems to suggest that it is precisely the 'mechanical' nature of this expedient that makes it superfluous. I essentially agree with Wittgenstein's characterization of proof as a sort of calculation, as a process during which one (often) does not bother "about sense or meaning",[5] but I will argue that it is precisely the fact of not bothering about sense or meaning and of applying rules that only deal with signs 'blindly' that makes this 'mechanical' way of reasoning particularly suitable to deliver surprises. So what I object to most of all is Wittgenstein's account of the *epistemology* of proof; in a Leibnizian vein, I shall argue that proof (and deductive reasoning in general) is in fact a crucial tool for the discovery of new facts, i.e. for information gain. What my argumentative strategy seeks to accomplish is to show that, even if we grant the other premises that Wittgenstein's argument is based on, it will still not follow that reasoning 'mechanically' will yield no surprises.

4 'Everyday life' reasoning and belief-bias

Decades of experimental research on the psychology of reasoning have shown that people typically do not reason according to the canons of deductive reasoning. After a long-lasting predominance of what could be described as a Piagetian paradigm, according to which reasoning in general (regardless of the circumstances) essentially follows the canons dictated by (presumably classical) logic, researchers started to conduct experiments whose results suggested significant discrepancy between 'everyday life' reasoning and the deductive canons; the locus classicus for this tradition is (Wason, 1966), where the famous Wason selection task was presented for the first time. The results presented by Wason indicated that subjects simply did not follow the basic logic of conditionals in order to solve the task proposed, namely to validate or invalidate a rule formulated as a conditional.

The 'fall' of the Piagetian paradigm has caused and still causes significant discomfort and turmoil, and many possible explanations have

---

[4] It can be argued that Wittgenstein is actually saying that proof is not superfluous in such complicated cases, but the fact that he uses the term 'facilitate' rather than 'ensure' or the like at least suggests that even in such cases proof is not essential.

[5] I will, however, highlight the importance of heuristic devices when formulating a proof at a later stage.

been proposed to account for the discrepancy between actual patterns of human reasoning as experimentally observed and the canons of deductive reasoning and of other presumably rational canons. Some have argued that there are fundamental problems with the experimental paradigm; others have concluded that people in general are bad reasoners and 'irrational'; yet others maintain that it is deductive reasoning instead that has no (natural) place in human cognition. (For an overview of this tradition, see (Evans, 2002).)

Among the many discrepancies studied since then, a group of reasoning mechanisms received the general name of 'cognitive biases'. The term 'bias' (in itself loaded with negative connotations) is used to describe a series of reasoning mechanisms that depart from the norm defined by the canons of 'rationality' theoretically defined (i.e. on the basis of deductive reasoning, rational choice theory etc.). Among these are confirmation bias and belief bias. Generally speaking, belief bias refers to the tendency we seem to have to endorse far more arguments as valid if their conclusions also accord with prior belief, and to reject arguments whose conclusions clash with prior belief. The concept was introduced in a 1983 paper by Evans et al. (Evans, Barston, & Pollard, 1983), to account for the results obtained in the following experiment: subjects were presented with syllogistic arguments and asked to indicate whether the conclusion necessarily followed from the premises given. In other words, subjects were asked to make an evaluation of the (logical) validity of the arguments in question. These were valid as well as invalid arguments, featuring plausible as well as implausible conclusions (in all four combinations). Some of the arguments presented to subjects were:

Valid-believable	Valid-unbelievable	Invalid-believable	Invalid-unbelievable
No police dogs are vicious.	No nutritional things are inexpensive.	No addictive things are inexpensive.	No millionaires are hard workers.
Some highly trained dogs are vicious.	Some vitamin tablets are inexpensive.	Some cigarettes are inexpensive.	Some rich people are hard workers.
Therefore, some highly trained dogs are not police dogs.	Therefore, some vitamin tablets are not nutritional.	Therefore, some addictive things are not cigarettes.	Therefore, some millionaires are not rich people.

The results once again clashed with the canons of deductive reasoning: subjects tended to endorse arguments with plausible conclusions (i.e. conclusions according with their prior beliefs) far more frequently than arguments with implausible conclusions, regardless of the argument's validity as such. Indeed, in the experiments *in*valid arguments with plausible conclusions are more often endorsed than valid arguments with *im*plausible conclusions, as the table below shows. This suggests that accordance of the conclusion with one's prior beliefs plays a more prominent role in the evaluation of the propriety of an argument than its 'logical' validity.

Percentage of conclusions accepted

*	Believable	Unbelievable
Valid	89	56
Invalid	71	10

Of course, these somewhat surprising results can be seen as a consequence of problems with the experimental setup; it is unclear, for example, how exactly subjects understand the concept of 'necessity' as presented in the task. Another hypothesis that has been put forward to account for the results is that subjects may be using a parsimonious heuristic mechanism: if the conclusion is already plausible, then there is no point in allocating cognitive and computational resources to evaluate the validity of the argument. (For a discussion of these hypotheses, see (Evans, Newstead, & Byrne, 1993, pp. 243–255). For a more recent article on belief-bias, see (Klauer, Musch, & Naumer, 2000).)

Be that as it may, belief-bias is a variation of a more general mechanism known as confirmation bias, i.e. a tendency to search for, interpret or remember information in a way that confirms preconceptions. People can reinforce their existing beliefs by selectively collecting new evidence, by interpreting evidence in a biased way or by selectively recalling information from memory. As studied in the literature, however, belief bias concerns confirmation bias specifically in the case of reasoning from premises to conclusion (and in particular concerning syllogistic arguments). True enough, the experiments set up to investigate belief bias typically consist in subjects being asked to endorse or reject previously formulated arguments rather than to draw conclusions themselves on the basis of given premises. Still, it would seem that the results at least suggest the more general pattern of belief bias as also operating when one is actively drawing inferences: one may typically avoid drawing conclusions from given (endorsed) premises if they happen to clash with one's prior beliefs, and may instead choose to draw conclusions that confirm these prior beliefs. These conclusions may actually 'follow' (validly) from the premises in question, but still the behavior of being selective and avoiding the counter-intuitive conclusions would already indicate a form of confirmation and belief-bias. Moreover, the results also suggest that there may be a tendency towards drawing 'conclusions' which do not actually follow from the given premises, but which are themselves plausible; this would amount to a fallacious form of seeking confirmation to one's prior beliefs (on this connection, see (Floridi, 2009).)

At any rate, what seems to be going on is that subjects let external information in the form of prior beliefs (in particular, their belief in the plausibility of the conclusion) 'contaminate' the reasoning process, whereas according to the cannons of deductive reasoning, only the information contained in the premises is to be taken into account. For my purposes here, this seems to be the main upshot of the belief bias experiments: we typically call upon external information when reasoning in everyday life, and for good reasons. In everyday life, to make use of deductive reasoning exclusively would be extremely counter-productive, as we typically do not dispose of sufficient information to draw the appropriate conclusions. We would 'freeze' most of the time, as the available information would underdetermine the conclusion(s) to be drawn.

Moreover, while the term 'bias' suggests that it is a mistake to seek confirmation to one's prior beliefs, in the literature there are also discussions on why it may be a perfectly 'reasonable' mechanism after all, in any case for use in practical situations, where there are constraints of time and cognitive resources (again, see (Evans et al., 1993, p. 243–255)). If one were to revise one's beliefs constantly, this would demand a significant allocation of cognitive resources, which would most likely provoke a situation of cognitive overload. We do, of course, perform belief revision regularly, but to suppress confirmation and belief bias completely would probably entail that we would be revising our beliefs more often than would be beneficial given our limited resources.

5 Deductive reasoning as a counterbalance to belief-bias

But there are contexts and circumstances where it is in fact not so advantageous to seek confirmation to previously held beliefs, i.e. to be conservative towards one's own prior doxastic commitments. This is the case in particular of scientific contexts, where the goal is precisely to uncover new information for as much as possible, and constraints of time and resources play a less significant role. So this may be yet another reason (besides the quest for the highest degree of certainty) why deductive reasoning is thought to be the quintessential form of reasoning in scientific contexts, that is, if it serves indeed as a counterbalance to belief bias, as I shall argue now.

In scientific contexts, the paradox of inference with which I started the paper manifests itself in the form of another tension, namely the tension between the certainty and indefeasibility sought after in science and the constant strive to reveal new facts, to produce new information. Typically, the higher the standards of certainty adhered to, the lower the amount of new information likely to be produced and accepted as legitimate. The use of deductive reasoning in science is often thought to be related to the high level of certainty it provides; but if I am correct in identifying deductive reasoning as a way to counterbalance confirmation and belief bias, then deductive reasoning may also play a positive role regarding information gain.

Let us now go back to Wittgenstein's characterization of proof in logic as a 'mechanical expedient'. The crucial features of this form of reasoning are disregard for 'sense or meaning' and the fact of using only rules that deal with signs. Wittgenstein is effectively saying that proof in logic is ultimately a form of calculation, of blind manipulation of signs by means of rules. While there is definitely more involved in derivations than mere calculation (more on this below), the process of 'de-semantification' (in S. Krämer's fitting terms in (Kramer, 2003)) that seems to be at least one important element of our use of formal languages when doing logic suggests that Wittgenstein is right in identifying a calculatory aspect in the act of formulating a proof. Applying the 'rules that deal with signs' is often done more efficiently precisely if one does not pause to think about the meaning of the expressions involved and instead treats them as 'meaningless signs'.

More importantly, this process of de-semantification also facilitates the suppression of contamination of prior beliefs in the reasoning process. Given that no ingenuity or insight is required to perform the application of the rules that deal with signs (even though some ingenuity may be required to decide *which* rules to apply — see below), it is effectively possible to avoid the interference of external information, in particular of prior beliefs.

How does this 'mechanical' way of reasoning relate to deductive reasoning more generally? The canons of deductive reasoning tell us that we must draw indefeasible inferences solely on the basis of the explicitly accepted premises, but they do not tell us how to accomplish this feat. Mechanical reasoning, or pure manipulation of signs, is one way of ensuring that no extra premises be allowed to sneak in during the reasoning process, and this is indeed one of Frege's crucial insights in the *Begriffsschrift*. While this feature of deductive reasoning is usually associated with the quest for the highest possible degree of certainty, my suggestion here is that it also contributes to the novelty of the conclusions to be drawn by means of purely logical reasoning, precisely because it blocks the possible interference of prior beliefs and thus our tendency to seek confirmation for them.

Even in logic, where one not only uses patterns of deductive reasoning to reason about a specific topic but also investigates these patterns as such, some expectations as to what is likely to be a theorem of a given theory are usually in place. And while not all results in logic are surprising, some are, and my suggestion here is that the 'mechanical way of reasoning' related to the manipulation of formal languages is at least partially responsible for the phenomenon of logicians overcoming their own confirmation bias.[6]

This being said, I must qualify that characterizing proofs in logic as a merely mechanical expedient does not tell us the whole story either. If nothing else, in most (interesting) logical systems the possibilities of combinations of 'the rules that deal with signs' amounts to a combinatorial explosion, and most of these combinations are utterly uninteresting and trivial. Proof in logic is not a merely mechanical expedient also because it takes insight to identify the promising paths, i.e. those leading to non-trivial results among the many possible com-

---

[6] Here is one example of a logical-mathematical result which very much surprised its own discoverer: the Paris-Harrington incompleteness theorems, which came as a total surprise to Jeff Paris (personal communication). For a broader discussion of surprising mathematical results, see http://rjlipton.wordpress.com/2009/09/27/surprises-in-mathematics-and-theory/.

binations. Indeed, even automated theorem-provers typically operate on the basis of heuristic mechanisms rather than on 'brute force' alone, i.e. mere combinatorial possibilities.

But ultimately, Wittgenstein's claim that there can be no surprises in logic stems from an undue disregard for the epistemic act of actually unpacking the information contained in the premises when drawing a (deductive) inference. Before the act of unpacking, the information contained in the premises is merely 'virtual information' (a term used by D'Agostino and Floridi), which must be actualized in order to become truly available, and this is exactly what the act of drawing an inference is able to accomplish. In a sense, the information is indeed contained in the premises, but in a sense it is not, i.e. in the sense of actually becoming available to us for future use. (It is not very different from telling a child who has just received a present that she need not open it, as the toy is already 'there' even though she cannot see it or play with it; the present must be unpacked in order to become 'real' for the child.)

As for surprises: the act of unpacking may reveal something that confirms what one already believes or at least suspects, but it may also reveal something that had not been taken notice of even though it was there all along, so to say, and which in fact clashes with one's prior beliefs. Deductive reasoning ought to be neutral with respect to prior beliefs not directly involved in a given deductive inference, and a certain degree of mechanization in one's reasoning can certainly contribute to this neutrality.

Catarina Dutilh Novaes
Philosophy Department and ILLC, University of Amsterdam
Nieuwe Doelenstraat 15, 1012 CP Amsterdam, The Netherlands
c.dutilhnovaes@uva.nl
http://staff.science.uva.nl/~dutilh/

References

Clark, M. (2002). *Paradoxes from a to z*. Cambridge: Camrbidge University Press.

Cohen, M., & Nagel, E. (1934). *An introduction to logic and scientific method*. London: Routledge and Kegan Paul.

D'Agostino, M., & Floridi, L. (2009). The enduring scandal of deduction. *Synthese, 167(2)*, 271-315.

Dummett, M. (1978). The justification of deduction. In *Truth and other enigmas* (p. 290-318). London: Duckworth.

Evans, J. (2002). Logic and human reasoning: an assessment of the deduction paradigm. *Psychological Bulletin, 128(6)*, 978-996.

Evans, J., Barston, J., & Pollard, P. (1983). On the conflict between logic and belief in syllogistic reasoning. *Memory and Cognition, 11*, 295-306.

Evans, J., Newstead, S., & Byrne, R. (1993). *Human reasoning.* East Sussex: L. Erlbaum.

Floridi, L. (2009). Logical fallacies as informational shortcuts. *Synthese, 167(2)*, 317-325.

Hintikka, J. (1973). *Logic, language-games and information.* Oxford: Clarendon.

Keynes, J. (1884). *Studies and exercises in formal logic.* London: Macmillian.

Klauer, K., Musch, J., & Naumer, B. (2000). On belief bias in syllogistic reasoning. *Psychological Review, 107(4)*, 852-884.

Kramer, S. (2003). Writing, notational iconicity, calculus: On writing as a cultural technique. *MLN, 118(3)*, 518-537.

Mill, J. (1843). *System of logic.* Cambridge: Cambridge University Press.

Primiero, G. (2008). *Information and knowledge.* Berlin: Springer.

Sequoiah-Grayson, S. (2008). The scandal of deduction. *Journal of Philosophical Logic, 37*, 67-94.

Wason, P. (1966). Reasoning. In B. Foss (Ed.), *New horizons in psychology* (p. 135151). Harmondsworth: Penguin.

Tenses and truth-conditions: a plea for if–then–else

Marie Duží*

1 Introduction

Sentences in the present, past and future tenses obviously have different truth-conditions. This fact has been observed by numerous logicians, and many variants of so-called temporal logic have been developed. These formal systems are mostly viewed as a special case of modal logic interpreted by means of Kripkean possible-world semantics. The term 'temporal logic' is broadly used to cover all approaches to the representation of the temporal dimension within a logical framework. More narrowly, it is also used to refer to a particular modal system of temporal propositional logic that Arthur Prior introduced in (Prior, 1957), (Prior, 1962) and (Prior, 1967) under the name '*tense logic*'.

The logical language of Prior's tense logic contains, in addition to the usual truth-functional operators, four modal operators whose intended meanings are:

P "It has at some time been the case that ..."
F "It will at some time be the case that ..."
H "It has always been the case that ..."
G "It will always be the case that ..."

P and F are known as the *weak tense operators*, while H and G are known as the *strong tense operators*. Prior developed a formal system of tense logic with axioms like

---

*This work has been supported by the Grant Agency of the Czech Republic, project No. GACR 401/07/0451, 'Semantization of Pragmatics', and by the internal grant agency of FEECS VSB–TU Ostrava — IGA 22/2009, 'Modelling, simulation and verification of software processes'.

$Gp \to Fp$ — "What will always be will be";

$G(p \to q) \to (Gp \to Gq)$ — "If p will always imply q then if p will always be the case, so will q";

$Fp \to FFp$ — "If it will be the case that p, it will be the case that it will be that p";

$\neg Fp \to F\neg Fp$ — "If it will never be that p then it will be that it will never be that p".

Similarly for the past operators P, H; e.g., $Hp \to Pp$, "What has always been has been". Subsequently, systems of temporal logic have been further developed by computer scientists, notably Zohar Manna and Amir Pnueli,[1] and widely used for formal verification of programs and for encoding temporal knowledge within artificial intelligence. In addition to Prior's future and past operators Manna and Pnueli have introduced modal operators like *Since* and *Until* that are provably more expressive than ordinary modal operators and are usually interpreted by (labelled) transition systems of program states that are pivotal to the operational semantics of programs.

These logics are undeniably simple, elegant and logically convenient. However, simplicity and convenience do not always go hand in hand with *logical adequacy*. Despite the great applicability of particular variants of tense logic in the semantics of programming languages, the systems just mentioned suffer a drawback when applied to the semantics of natural language. The drawback is their inability to adequately analyse sentences indicating a 'point of reference' referring to the interval when the sentence was or will be true. Such sentences come attached with a *presupposition* under which a sentence is true or false. To illustrate the problem, consider the sentences

"Tom is sick".

"Tom has been sick".

"Tom was sick throughout October 2009".

"Tom will be sick the whole day on April 1st, 2010".

The first two sentences do not come with a presupposition. They ascribe to Tom the property of being sick and of having been sick,

---

[1] See (Manna & Pnueli, 1992), (Manna & Pnueli, 1995).

respectively. They are true or false according as Tom has the relevant property. However, the truth-condition of the third sentence depends not only on whether Tom has the property of being sick throughout October 2009, but also on the time at which the sentence is evaluated. If T is the time of evaluation then the truth-conditions are specified as follows:

If $T \leq$ October 31, 2009, 24:00, *then* **no truth-value**
else **True** or **False** according as Tom was sick
at all times during October 2009.

Similarly, the fourth sentence comes attached with a presupposition that the time T of evaluation comes before April 1st, 2010.

Our analysis must respect these truth-conditions. To this end we apply the rich system of Tichý's Transparent Intensional Logic (TIL).[2] Tichý put forward his solution in (Tichý, 1980). However, this solution is difficult to understand, because Tichý applies the *singulariser* function to a singleton typed as containing a truth-value in order to make the set fail to deliver a truth-value in case the associated presupposition is not satisfied. Tichý's analysis is analogous to what the computer scientist would call an *imperative* rather than *declarative* analysis. The downside to an imperative analysis is that it may conceal flaws that rear their head only when the analysis is applied to extreme situations. Yet there is an elegant alternative that makes use of the 'if–then–else' connective, which I am going to introduce in Section 3 of this paper.[3]

There has been much dispute over the semantics of 'if–then–else' in the logic of computer science. It cannot be adequately analyzed by means of material implication. The reason is this. The application of *if–then–else* to a condition P and two formulae F_1 and F_2 is not improper (by failing to provide a truth-value) even when F_2 is improper whenever P is true or when F_1 is improper whenever P is false. However, the regimentation of "If P then F_1 else F_2" in propositional logic that takes the form $[(P \supset F_1) \wedge (\neg P \supset F_2)]$ is improper by failing to produce a truth-value whenever F_1 or F_2 is improper regardless the condition P. Thus it is often said that *if–then–else* is a non-strict

---
[2]See (Tichý, 1988).
[3]I am grateful to Nikola Ciprich for drawing my attention to this option.

function that does not behave in compliance with the compositionality principle. Yet there is no cogent reason to settle for non-strictness.

In what follows I am going to show that the *procedural* semantics of TIL enables us to specify a strict definition of *if–then–else* that meets the compositionality constraint. The definition of "If P then F_1 else F_2" is a procedure that decomposes into two phases. First, on the basis of the condition P, select one of F_1, F_2 as the procedure to be executed. Second, execute the selected procedure. Thus, for instance, if P is true then F_1 is executed rather than F_2, and the possible improperness of F_2 does not matter. After setting out the definition, I specify a general schema in which to couch the analysis of sentences that come attached with a presupposition, in particular sentences in the past and future tenses.

2 Method of analysis

TIL operates with a single *procedural semantics* for all kinds of logical-semantic context, whether extensional, intensional or hyper-intensional.[4] It means that it explicates the meaning of an expression as an abstract *procedure* encoded by the expression. Such procedures are rigorously defined as TIL *constructions* and we assign them to expressions as their *context-invariant* meanings. From the formal point of view, TIL is a hyper-intensional, partial, typed λ-calculus. Hyper-intensional, because the terms of the TIL formal language in which constructions are encoded are interpreted as *procedures* (generalized algorithms) rather than their *products*; partial, because the primitive notion of TIL is function understood as a partial mapping that assigns to each element of its domain *at most one* element of its range; and typed, because all the entities of TIL ontology, including constructions, receive a type.

Intuitively, construction C is a *procedure* (a generalised algorithm). Constructions are *structured* in the following way. Each construction C consists of sub-constructions (*constituents*), each of which needs to be executed when executing C. Specification of a construction can be viewed as an instruction on how to proceed in order to obtain the output entity given some input entities. In this way a construction

---

[4] In this section, the philosophy and basic notions of TIL are only briefly summarized. For details, see (Duží, Jespersen, & Materna, forthcoming).

constructs a function understood as a mapping from the set of input entities into a set of output entities.

There are two kinds of constructions, atomic and compound (molecular). Atomic constructions (*Variables* and *Trivializations*) do not contain any other constituent but themselves; they specify objects (of any type) on which compound constructions operate. The *variables* $x, y, p, q, \ldots$, construct objects dependently on a valuation; they v-construct. The *Trivialisation* of an object X (of any type, even a construction), in symbols ^0X, constructs simply X without the mediation of any other construction. *Compound* constructions, which consist of other constituents as well, are *Composition* and *Closure*. The *Composition* $[FA_1 \ldots A_n]$ is the operation of functional application. It v-constructs the value of the function f (*valuation-*, or v-, -constructed by F) at a tuple argument A (v-constructed by $A_1, \ldots, A_n$), if the function f is defined at A, otherwise the Composition is v-*improper*, i.e., it *fails* to v-construct anything.[5] The *Closure* $[\lambda x_1 \ldots x_n F]$ spells out the instruction to v-construct a function by abstracting over the values of the variables $x_1, \ldots, x_n$ in the ordinary manner of the λ-calculi.[6] Finally, higher-order constructions can be used twice over as constituents of composite constructions. This is achieved by a fifth construction called *Double Execution*, ^2X, that behaves as follows: If X v-constructs a construction Y, and Y v-constructs an entity Z then ^2X v-constructs Z; otherwise ^2X is v-*improper*, failing as it does to v-construct anything.

TIL constructions, as well as the entities they construct, all receive a type. The formal ontology of TIL is bi-dimensional; one dimension is made up of constructions, the other dimension encompasses non-constructions. On the ground level of the type hierarchy, there are non-constructional entities unstructured from the algorithmic point of view belonging to a *type of order* 1. Given a so-called *epistemic* (or *objectual*) base of *atomic types* (o-truth values, ι-individuals, τ-time moments/real numbers, ω-possible worlds), the induction rule for forming functional types is applied: where $\alpha, \beta_1, \ldots, \beta_n$ are types

---

[5] We treat functions as *partial mappings*, i.e., set-theoretical objects, unlike the *constructions* of functions.

[6] Comparison with programming languages might be helpful: λ-Closure corresponds to a declaration of a procedure F with formal parameters $x_1, \ldots, x_n$; Composition corresponds to calling the procedure F with actual values $A_1, \ldots, A_n$ of parameters.

of order 1, the set of partial mappings from $\beta_1 \times \cdots \times \beta_n$ to α, denoted '$\alpha\beta_1\ldots\beta_n$', is a type of order 1 as well. Constructions that construct entities of order 1 are *constructions of order* 1. They belong to a *type of order* 2, denoted '$\star_1$'. The type $\star_1$ together with atomic types of order 1 serve as a base for the induction rule: any collection of partial mappings, type $(\alpha\beta_1\ldots\beta_n)$, involving $\star_1$ in their domain or range is a *type of order* 2. Constructions belonging to a type $\star_2$ that v-construct entities of order 1 or 2, and partial mappings involving such constructions, belong to a *type of order* 3. And so on *ad infinitum*.

The sense of an empirical expression is a *hyperintension*, i.e., a construction that produces a (possible world) *(α-)intension*; α-intensions are members of type $(\alpha\omega)$, i.e., functions from possible worlds to an arbitrary type α. On the other hand, *(α-)extensions* are members of a type α, where α is not equal to $(\beta\omega)$ for any β, i.e., extensions are not functions whose domain are possible worlds. Intensions are frequently functions of a type $((\alpha\tau)\omega)$, i.e., functions from possible worlds to *chronologies* of the type α (in symbols: $\alpha_{\tau\omega}$), where a chronology is a function of type $(\alpha\tau)$.

Some important kinds of intensions are:

- *Propositions, type* $o_{\tau\omega}$ They are denoted by empirical sentences.

- *Properties of members of a type α, or simply α-properties, type* $(o\alpha)_{\tau\omega}$.[7] General terms, some substantives, intransitive verbs ('student', 'walks') denote properties, mostly of individuals.

- *Relations-in-intension, type* $(o\beta_1\ldots\beta_m)_{\tau\omega}$. For example transitive empirical verbs ('like', 'worship'), also attitudinal verbs denote these relations.

- α-*roles*, also α-*offices, type* $\alpha_{\tau\omega}$, *where* $\alpha \neq (o\beta)$; frequently $\iota_{\tau\omega}$. They are often denoted by concatenation of a superlative and a noun ('the highest mountain').

Notational conventions. An object A of a type α is denoted 'A/α'. That a construction $C/\star_n$ v-constructs an object of type α is denoted '$C \to_v \alpha$'. We use variables $w, w_1, \ldots$ as v-constructing elements of

---

[7]We model α-sets and $(\alpha_1\ldots\alpha_n)$-relations by their characteristic functions of type $(o\alpha)$, $(o\alpha_1\ldots\alpha_n)$, respectively. Thus an α-property is an empirical function that dependently on states-of-affairs $(\tau\omega)$ picks-up a set of α-individuals, the population of the property.

type ω (possible worlds), and $t, t_1, \ldots$ as v-constructing elements of type τ (times). If $C \rightarrow_v \alpha_{\tau\omega}$ v-constructs an α-intension, the frequently used Composition of the form $[[Cw]t]$, the intensional descent of the α-intension, is abbreviated 'C_{wt}'.

Quantifiers, $\forall^\alpha$ (the general one) and $\exists^\alpha$ (the existential one), are of types $(o(o\alpha))$, i.e., sets of sets of α-objects. $[^0\forall^\alpha \lambda x A]$ v-constructs the truth-value **T** iff $[\lambda x A]$ v-constructs the whole type α, otherwise **F**; $[^0\exists^\alpha \lambda x A]$ v-constructs **T** iff $[\lambda x A]$ v-constructs a non-empty subset of the type α, otherwise **F**. We write '$\forall x A$', '$\exists x A$' instead of $[^0\forall^\alpha \lambda x s A]$, $[^0\exists^\alpha \lambda x A]$, respectively, when no confusion can arise. In the effort of easier reading we will also use an infix notation without trivialisation when using constructions of truth-value functions of type (ooo), i.e., $\land$ (conjunction), $\lor$ (disjunction), $\supset$ (implication), $\equiv$ (equivalence), and negation ($\neg$) of type (oo), and when using constructions of common relations like identities, less than ($<$), greater than ($>$), etc.

We invariably furnish expressions with their procedural structured meanings, which are explicated as TIL constructions. The analysis of a sentence thus consists in discovering the logical construction encoded by a given sentence. The TIL compositional *method of analysis* is driven by Carnap's *principle of subject matter*, which says, roughly, that only those entities that receive mention in a sentence can serve as constituents of the sentence meaning.[8] The method decomposes into three phases:

1. *Type-theoretical analysis*, i.e., assigning types to the objects that receive mention in the analysed sentence.

2. *Synthesis*, i.e., combining the constructions of the objects *ad* 1. in order to construct the proposition of type $o_{\tau\omega}$ denoted by the whole sentence.

3. *Type-theoretical checking*.

To illustrate the method let us analyse the sentence "Tom is sick".
Ad 1. Tom/ι; Sick/$(o\iota)_{\tau\omega}$: the property of individuals (of being sick).
Ad 2. Since we aim at discovering the *literal* analysis of the sentence, objects denoted by semantically simple expressions 'Tom', 'is sick' are constructed by their Trivialisations: ^{0}Tom, ^{0}Sick. Now we need to predicate the property *Sick* of the individual *Tom*. But we

---
[8]See (Carnap, 1947, § 24.2, § 26).

cannot apply the property directly to *Tom*. The individual *Tom* is not a type-theoretically proper object to serve as an argument of a property, because property is an *intension*. It must be extensionalized first, i.e., applied to a given world w and time t of evaluation, and only then applied to Tom. This is achieved by three Compositions [^{0}Sick$w$] $\to_v$ (($o\iota$)τ), [[^{0}Sick$w$]t] $\to_v$ ($o\iota$), abbreviated as ^{0}Sick$_{wt}$, and finally [[[^{0}Sick$w$]t] ^{0}Tom] $\to_v$ o, or [^{0}Sick$_{wt}$ ^{0}Tom] for short.[9] Since we are going to construct a proposition, i.e., an intension, we must abstract from values of w and t:

$$\lambda w \lambda t [^0\text{Sick}_{wt} \, ^0\text{Tom}].$$

This construction is the literal analysis of the sentence "Tom is sick".

Ad 3. *Type-theoretical checking:*

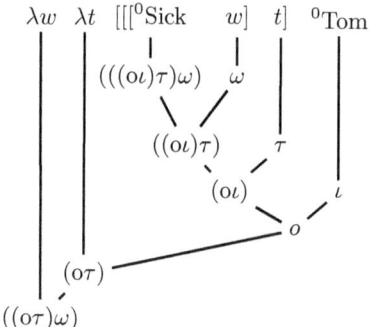

Figure 1.

The role of Trivialisation and empirical parameters, that is modal $w \to \omega$ and temporal $t \to \tau$ ones, can be elucidated as follows. When evaluating the truth-condition of the sentence, then the upper index '0' serves as a marker of the primitive concept that supplies an object to be operated on. The lower index '$_{wt}$' can be understood as an instruction to execute an *empirical inquiry (search)* in order to check whether the object in question (here Tom) satisfies the condition specified by the intension (here sickness).

So much for the TIL method of analysis. Now I am going to apply this method to analyze sentences in past and future.

---

[9] For details on predicating properties of individuals, see (Jespersen, 2008).

3 Sentences in past and future tenses

Consider the sentence in present perfect

$$\text{"Tom has been sick"}. \tag{1}$$

It informs us that Tom has the property of being sick not only in the present time t of evaluation but also in an interval that runs from the past up to t (and possibly beyond). Thus the analysis of the sentence comes down to the Closure

$$\lambda w \lambda t \exists t_1 [\forall t_2 [t_1 < t_2 \leq t] \supset [{}^0\text{Sick}_{wt_2}\,{}^0\text{Tom}]]. \tag{1'}$$

A similar sentence in simple past, which is "Tom was sick", seems to be incomplete, because one is tempted to ask "When was Tom sick?" This is because sentences in past should contain an indication of *when* something happened, for instance

$$\text{"Tom was sick throughout October 2009"}. \tag{2}$$

As mentioned above, such a sentence not only entails but also *presupposes* that the time t of evaluation comes after October 2009. The difference between presupposition and mere entailment can be schematically demonstrated as follows:

(i) P is a *presupposition* of S, iff: $(S \models P)$ and $(non\text{-}S \models P)$

   Corollary: If $non\text{-}P$ then neither S nor $non\text{-}S$ is true, i.e., S does not have a truth-value.

(ii) Mere entailment: $(S \models P)$ and neither $non\text{-}S \models P)$ nor $(non\text{-}S \models non\text{-}P)$

More precisely, the entailment relation $\models$ obtains between hyper-propositions P, S, i.e., the meaning of P is entailed or presupposed by the meaning of S. And since we work with *partial* functions, we can smoothly analyse sentences associated with a presupposition. If the proposition constructed by P does not take the truth-value **T** at a given $\langle w, t \rangle$-pair, then the proposition constructed by S has a truth-value gap at this $\langle w, t \rangle$.

Denoting the interval October 2009 by 'Oct09', the presupposition of (2) is

$$\lambda w \lambda t \forall t_1 [[^0\text{Oct09}\, t_1] \supset [t_1 < t]].$$

Gloss: In any world w at any time t it holds for all times $t_1 \to \tau$ belonging to the interval Oct09/($o\tau$) that $t_1 < t$. In other words, the entire month October 2009 precedes the time t.

Now the schematic analysis of (2) is this:

$$\lambda w \lambda t \; \text{If} \; \forall t_1 [[^0\text{Oct09}\, t_1] \supset [t_1 < t]] \\ \text{then} \; \forall t'[[^0\text{Oct09}\, t'] \supset [^0\text{Sick}_{wt'}\, {}^0\text{Tom}]] \; \text{else Fail}. \qquad (2^s)$$

To complete the analysis, we must define the *If–then–else* function. Here is how. The instruction encoded by "If $P(\to o)$ then $C(\to \alpha)$, else $D(\to \alpha)$" behaves as follows:

(a) If P v-constructs **T** then execute C (and return the result of type α, provided C is not v-improper).

(b) If P v-constructs **F** then execute D (and return the result of type α, provided D is not v-improper).

(c) If P is v-improper then fail to produce the result.

Hence, *if–then–else* is seen to be a function of type $(\alpha\, o\star_n \star_n)$, and its definition decomposes into two phases.

First, select a construction to be executed on the basis of a specific condition P. The choice between C and D comes down to this Composition:

$$[^0\text{The\_only}\, \lambda c[[P \supset [c = {}^0C]] \wedge [\neg P \supset [c = {}^0D]]]].$$

Types: $P \to_v o$ v-constructs the condition of the choice between the execution of C or D, $C/\star_n$, $D/\star_n$; $c \to_v \star_n$; The_only/($\star_n(\star_n)$): the singularizer function that associates a singleton set of constructions with the only construction that is an element of this singleton, and is otherwise (i.e., if the set is empty or many-valued) undefined. If P v-constructs **T** then the variable c v-constructs the construction C, and if P v-constructs **F** then the variable c v-constructs the construction D. In either case, the set constructed by

$$\lambda c[[P \supset [c = {}^0C]] \wedge [\neg P \supset [c = {}^0D]]]$$

is a singleton and the singularizer The_only returns as its value either the construction C or the construction D. Note that in this phase constructions C and D are not constituents to be executed; rather they are mere objects to be supplied by the variable c. This is to say that without *hyperintensional* approach we would not be able to define the function *If–then–else*.

Second, the selected construction is executed; therefore, Double Execution must be applied:

$$^2[^0\text{The\_only}\,\lambda c[[P \supset [c = {}^0C]] \wedge [\neg P \supset [c = {}^0D]]]].$$

As a special case of P being a presupposition, *no* construction D is to be selected whenever P is not satisfied. Thus the analysis of

"If (presupposition) P then $C \to o$

else *Fail* (to produce a truth-value)"

comes down to the Double Execution

$$^2[^0\text{The\_only}\,\lambda c\,[[P \supset [c = {}^0C]] \wedge [\neg P \supset {}^0\mathbf{F}]]].$$

Gloss: If $\neg P$ v-constructs $\mathbf{T}$ then $[\neg P \supset {}^0\mathbf{F}]$ v-constructs $\mathbf{F}$ and the set v-constructed by the Closure $\lambda c[[P \supset [c = {}^0C]] \wedge [\neg P \supset {}^0\mathbf{F}]]$ is empty. Thus the singulariser The_only does not return any construction and the Double Execution does not obtain an argument to execute; hence it is v-improper, that is fails to produce a truth-value.

Back to the analysis of (2). Applying this schematic definition to the construction (2^s), that is, substituting $\forall t_1[[^0\text{Oct09}\,t_1] \supset [t_1 < t]]$ for P and $[\forall t'[[^0\text{Oct09}\,t'] \supset [^0\text{Sick}_{wt'}\,{}^0\text{Tom}]]]$ for C, we obtain the final analysis of the sentence (2):

$$\lambda w \lambda t\,{}^2[^0\text{The\_only}\,\lambda c[[\forall t_1[[^0\text{Oct09}\,t_1] \supset [t_1 < t]] \supset$$
$$[c = {}^0[\forall t'[[^0\text{Oct09}\,t'] \supset [^0\text{Sick}_{wt'}\,{}^0\text{Tom}]]]]] \wedge \qquad (2^*)$$
$$\wedge\,[\exists t_1[[^0\text{Oct09}\,t_1] \wedge [t_1 \geq t]] \supset {}^0\mathbf{F}]]].$$

Since such an analysis is not easy to read and the *If–then–else* function has been defined, in what follows I will use the schematic analysis of the form

"$\lambda w \lambda t\,\text{If}\,P_{wt}\,\text{then}\,S_{wt}\,\text{else}\,\text{Fail}$"

rather than the full-fledged

"$\lambda w \lambda t\ ^2[^0\text{The\_only}\lambda c[[P_{wt} \supset [c = {}^0[S_{wt}]]] \wedge [\neg P_{wt} \supset {}^0\mathbf{F}]]]$".

The sentences in past often indicate as a reference point not only an interval when something happened but also a frequency of it, like once, twice, often, or throughout (as is the case of (2)). To adduce another example, consider the sentence

"Tom was sick (just) twice in October 2009". (3)

The presupposition of (3) is again the proposition that the entire month October 2009 precedes time t of evaluation: $\lambda w \lambda t \forall t_1 [[^0\text{Oct09}\, t_1] \supset [t_1 < t]]$. The schematic analysis of (3) comes down to:

$\lambda w \lambda t$ If $\forall t_1 [[^0\text{Oct09}\, t_1] \supset [t_1 < t]]$
then $[[^0\text{Twice}_w \lambda w \lambda t [^0\text{Sick}_{wt}\, ^0\text{Tom}]]\, ^0\text{Oct09}]$ else Fail. (3^s)

The frequency modifier *Twice* denotes a world-dependent function that takes a proposition $p \to o_{\tau w}$ to the class of those intervals $d \to (o\tau)$ which are contained in the chronology of p (i.e. $p_w \to (o\tau)$). This class of intervals d that have a non-empty intersection with a reference interval c is of cardinality two. Thus the application of *Twice* of type $((o(o\tau))\, o_{\tau w})_w$ to a proposition p and reference interval c comes down to this Composition:

$[[^0\text{Twice}_w p]c] = [^0\text{Card}\lambda d[\forall t[[dt] \supset p_{wt}] \wedge \exists t[[dt] \wedge [ct]]] = {}^0 2]$.

In our case the interval c is Oct09 and the proposition p is

$\lambda w \lambda t [^0\text{Sick}_{wt}\, ^0\text{Tom}]$,

and we have

$[[^0\text{Twice}_w \lambda w \lambda t [^0\text{Sick}_{wt}\, ^0\text{Tom}]]\, ^0\text{Oct09}] =$
$[^0\text{Card}\lambda d[\forall t[[dt] \supset [^0\text{Sick}_{wt}\, ^0\text{Tom}]] \wedge \exists t[[dt] \wedge [^0\text{Oct09}\, t]]] = {}^0 2]$.

Thus we can refine the analysis (3^s) like this:

$\lambda w \lambda t$ If $\forall t_1 [[^0\text{Oct09}\, t_1] \supset [t_1 < t]]$ then
$[^0\text{Card}\lambda d[\forall t[[dt] \supset [^0\text{Sick}_{wt}\, ^0\text{Tom}]] \wedge \exists t[[dt] \wedge [^0\text{Oct09}\, t]]] = {}^0 2]$ (3*)
else Fail.

Tenses and truth-conditions: a plea for if–then–else   75

Our resources up to now make it possible to define a general schema of the analysis of a sentence S in past tense with a reference interval In_Time/$(o\tau)$ and a modifier Frequency/$((o(o\tau))\,o_{\tau w})_w$. Let *Past* be a time-dependent function that takes a class of *o*-chronologies (the intervals in which a given proposition is true) together with an (implicit or explicit) reference interval and returns **T**, **F** or no value, according as the interval serving as point of reference belongs to the respective class of *o*-chronologies *and* precedes the time T at which the proposition denoted by the sentence is being evaluated. Thus, *Past* is typed as $((o(o(o\tau))(o\tau))\tau)$. Let $\leq_\tau /(o(o\tau)\tau)$ mean that the reference interval In_Time is prior to time t. Then the **general schema of sentences in past** is:

$$\lambda w \lambda t [{}^0 \text{Past}_\tau [{}^0 \text{Frequency}_w S] \, {}^0 \text{In\_Time}] =$$
$$\lambda w \lambda t \, \text{If} \, [{}^0 \text{In\_Time} \leq_\tau t] \, \text{then} \, [[{}^0 \text{Frequency}_w S]{}^0 \text{In\_Time}] \, \text{else Fail.}$$

For instance, our sentence (2) receives the literal analysis

$$\lambda w \lambda t [{}^0 \text{Past}_t [{}^0 \text{Throughout}_w \lambda w \lambda t [{}^0 \text{Sick}_{wt} \, {}^0 \text{Tom}]] \, {}^0 \text{Oct09}]$$

and the sentence

"Tom was sick at least once before October 2009"   (4)

is analysed by the Closure

$$\lambda w \lambda t [{}^0 \text{Past}_t [{}^0 \text{At\_least\_once}_w \lambda w \lambda t [{}^0 \text{Sick}_{wt} \, {}^0 \text{Tom}]]$$
$$\lambda t' [{}^0 \text{Before}t' \, {}^0 \text{Oct09}]] =$$

$$= \lambda w \lambda t \, \text{If} \, [\lambda t' [{}^0 \text{Before}t' \, {}^0 \text{Oct09}] \leq_\tau t]$$
$$\text{then} \, [[{}^0 \text{At\_least\_once}_w \lambda w \lambda t [{}^0 \text{Sick}_{wt} \, {}^0 \text{Tom}]] \lambda t' [{}^0 \text{Before}t' \, {}^0 \text{Oct09}]]$$
$$\text{else Fail.}$$

Here the point of reference is specified as *any time before Oct09*. To analyse 'before October 2009', we have to define the type of the object denoted by 'before'. Given a time t' and a τ-class c, the time t' is prior to c if t' is prior to every element of c. Thus Before/$(o\tau(o\tau))$ receives the definition ${}^0 \text{Before} = \lambda t' c [\forall t [ct] \supset [t' < t]]$, and 'before October

2009' expresses the Closure $\lambda t'[^0\text{Before}\, t'\, {}^0\text{Oct09}]$ which is equivalent to $\lambda t'[\forall t[{}^0\text{Oct09}\, t] \supset [t' < t]]$.

The definition of At_least_once$/((o(o\tau))\, o_{\tau w})_\omega$ is easy:

$${}^0\text{At\_least\_once} = \lambda w \lambda p \lambda c \exists t[[ct] \wedge p_{wt}].$$

The truth-condition is that a proposition p be true at least once in a world w in an interval c if there is at least one time t in c at which p is true in w. Thus the Composition $[{}^0\text{At\_least\_once}_w \lambda w \lambda t[{}^0\text{Sick}_{wt}\, {}^0\text{Tom}]]$ v-constructs the class $S/(o(o\tau))$ of intervals in which Tom is sick at least once.

The **general schema of sentences in future** is similar to the analytic schema of sentences in past:

$$\lambda w \lambda t[{}^0\text{Future}_t[{}^0\text{Frequency}_w\, S]\, {}^0\text{In\_Time}] =$$
$$\lambda w \lambda t \,\text{If}\, [{}^0\text{In\_Time} \geq_\tau t] \,\text{then}\, [[{}^0\text{Frequency}_w S]\, {}^0\text{In\_Time}] \,\text{else}\, Fail.$$

Here $\geq_\tau$ means that the reference interval In_Time comes after time t, Future receives the same type as Past, that is $((o(o(o\tau))(o\tau))\tau)$. For instance, the sentence

"Tom will be sick the whole day on April 1st 2010" (5)

expresses as its sense (April 1$/(o\tau)$: the day April 1st, 2010)

$$\lambda w \lambda t[{}^0\text{Future}_t[{}^0\text{The\_whole}_w \lambda w \lambda t[{}^0\text{Sick}_{wt}\, {}^0\text{Tom}]]\, {}^0\text{April 1}] \quad (5^s)$$

which is equivalent to

$$\lambda w \lambda t \,\text{If}\, [{}^0\text{April 1} \geq_\tau t]$$
$$\text{then}\, [[{}^0\text{The\_whole}_w \lambda w \lambda t[{}^0\text{Sick}_{wt}\, {}^0\text{Tom}]]\, {}^0\text{April 1}]$$
$$\text{else Fail.}$$

This analysis can be refined to this Closure:

$$\lambda w \lambda t \,\text{If}\, \forall t_1[[{}^0\text{April 1}\, t_1] \supset [t_1 > t]]$$
$$\text{then}\, \forall t'[[{}^0\text{April 1}\, t'] \supset [{}^0\text{Sick}_{wt'}\, {}^0\text{Tom}]]$$
$$\text{else Fail.}$$

4 Topic-focus ambiguities

Now I am going to heed the ambiguities pivoted on the difference between topic and focus articulation.[10] As an example, consider the sentence

$$\text{"All Tom's children were sick last week"}. \tag{6}$$

There are two non-equivalent readings of this sentence. To illustrate, imagine two scenarios.

(i) The sentence is an answer to the question "What about Tom's children"? Then 'Tom's children' is the topic of the sentence and to each of these children the property of being sick last week (the focus) is ascribed. In such a situation the sentence not only entails but also *presupposes* that Tom has children *in the present time*.

(ii) Another possible scenario is this. The question is "What was going on *last week*"? And the answer, "Oh, all Tom's children were sick *last week*". In this situation 'last week' is the topic and the sentence *only entails* but does not presuppose that Tom had children *last week*.

Since the sentence is ambiguous, we are actually going to analyse *two* non-equivalent sentences, which might be paraphrased as follows:

"Each of present Tom's children was sick (throughout) last week". (6i)

"(Throughout) last week each of the children Tom had was sick". (6ii)

Let Last_week/$((o\tau)\tau)$ be the function that associates a given time t with an interval that is last week with respect to t. The analyses come down to these Closures:

---

[10] For a linguistic analysis of this difference see (Hajičová, 2008).

$\lambda w\lambda t$ If [^{0}Has$_{wt}$ ^{0}Tom ^{0}Children]
then $\forall t^*$[[[^{0}Last_week$t$]t^*] $\supset$ [[^{0}All[^{0}Children_of$_{wt}$ ^{0}Tom]] ^{0}Sick$_{wt^*}$]]
else Fail.

(6i*)

$\lambda w\lambda t\forall t^*$[[[[^{0}Last_week$t$]$t*$] $\supset$ [^{0}Has$_{wt^*}$ ^{0}Tom ^{0}Children]]$\wedge$
[[^{0}Has$_{wt^*}$ ^{0}Tom ^{0}Children] $\supset$ [[^{0}All [^{0}Children_of$_{wt^*}$ ^{0}Tom]] ^{0}Sick$_{wt^*}$]]]

(6ii*)

Types: All/$((o(o\iota))(o\iota))$: the restricted quantifier that associates a set M of individuals with the set of supersets of M. Has/$(o\iota(o\iota)_{\tau\omega})_{\tau\omega}$: the relation-in-intension between an individual and a property (of having instances of the property); Children/$(o\iota)_{\tau\omega}$; Children_of/$((o\iota)\iota)_{\tau\omega}$;

Note that indeed (6ii*) only entails that Tom had children in all times t^* belonging to the last week, but does not presuppose it. This is because (6ii*) constructs a proposition that takes value **F** in those $\langle w, t\rangle$-pairs where either Tom did not have children in times t^* or Tom had children at that time but some of them were not sick.

5 Concluding remarks

In this paper I demonstrated the method of analysis of sentences in past and future. Moreover, I also presented the general analytic schema for sentences that come associated with a presupposition. To this end I utilized a strict definition of the *If–then–else* function that complies with the compositionality constraint. Last but not least, the semantic character of the ambivalence concerning the topic-focus articulation of sentences was analysed.

Logical analysis cannot disambiguate any sentence, because it presupposes full linguistic competence. Yet, our fine-grained method can contribute to a language disambiguation by making these hidden features *explicit* and *logically tractable*. In case there are more non-equivalent senses of a sentence we furnish the sentence with different TIL constructions. Having a formal fine-grained encoding of a sense, we can then *infer the relevant consequences*.

To sum up, I am convinced that if any logic can serve to solve such hard problems like fine-grained analysis of tenses, topic-focus ambiguities, and many others that natural language can produce, then it must be a logic with *hyper-intensional (most probably procedural) semantics*, such as TIL.

Marie Duží
Department of Computer Science
VSB-Technical University Ostrava
17. listopadu 15
708 33 Ostrava, Czech Republic
marie.duzi@vsb.cz
http://www.cs.vsb.cz/duzi

References

Carnap, R. (1947). *Meaning and Necessity*. Chicago: Chicago University Press.

Duží, M., Jespersen, B., & Materna, P. (forthcomming). *Procedural Semantics for Hyperintensional Logic; Foundations and Applications of Transparent Intensional Logic*. Berlin: Springer.

Hajičová, E. (2008). What we are talking about and what we are saying about it. *Computational Linguistics and Intelligent Text Processing*, *4919*, 241-262.

Jespersen, B. (2008). Predication and extensionalization. *Journal of Philosophical Logic*, *37*, 479–499.

Manna, Z., & Pnueli, A. (1992). *The Temporal Logic of Reactive and Concurrent Systems: Specification*. New York: Springer-Verlag.

Manna, Z., & Pnueli, A. (1995). *Temporal Verification of Reactive Systems: Safety*. New York: Springer-Verlag.

Prior, A. N. (1957). *Time and Modality*. Oxford: Oxford University Press.

Prior, A. N. (1962). Tense logic and the continuity of time. *Studia Logica*, *13*, 133–148.

Prior, A. N. (1967). *Past, Present and Future*. Oxford: Clarendon Press.

Tichý, P. (1980). The logic of temporal discourse. *Linguistics and Philosophy*, *3*, 343–369. (reprinted in (Tichý, 2004))

Tichý, P. (1988). *The Foundations of Frege's Logic*. Berlin: New York: De Gruyter.

Tichý, P. (2004). *Collected Papers in Logic and Philosophy* (V. Svoboda, B. Jespersen, & C. Cheyne, Eds.). Prague: Filosofia, Czech Academy of Sciences and Dunedin: University of Otago Press.

Some Critical Remarks on Incompatibility Semantics

Christian G. Fermüller*

1 Introduction

In his fifth Locke Lecture (2006)[1] Robert Brandom has presented a new type of semantics for propositional classical and modal logic ('incompatibility semantics') that is embedded in his quite general programme ('inferential pragmatism') addressing major challenges to analytic philosophy. Inferential pragmatism is an important, comprehensive, and widely discussed contribution to contemporary philosophy. This is not the place to comment on this programme in general. Rather, we want to draw attention to a particular problem with the semantic framework suggested by Brandom, arising from a misinterpretation of an allegedly central technical result. This misunderstanding has strong repercussions on the philosophical significance of incompatibility semantics.

The main features that Brandom ascribes to his incompatibility semantics can be briefly summarized as follows:

1. it is based on the notion of *material incompatibility* of (interpreted) sentences, rather than on their truth;

2. it is *strongly intensional*, treating conjunction and, in particular, negation on a par with the modal operator 'necessarily';

3. it is *holistic* and *non-compositional* in the sense that the meaning of a given compound formula (sentence) F is not determined by the semantic interpretants of the subformulas and connectives occurring in F;

---

*This work is supported by Eurocores-ESF/FWF grant I143-G15 (LogICCC-LoMoReVI).

[1] The lectures are accessible at http://www.pitt.edu/~rbrandom/ and published as Chapter 5 of (Brandom, 2008). I first learned about this exciting endeavor from a very stimulating invited talk of Brandom at LOGICA2008.

4. it nevertheless enjoys *recursive projectibility*, i.e., incompatibilities between logically complex formulas are determined by the incompatibilities between formulas that are less complex;

5. it *refutes* the claim that holistic semantics *cannot account for* the projectibility and systematicity of a language, and hence also not for its *learnability*.

As indicated, Brandom's presentation of incompatibility semantics is not only motivated from a broader philosophical perspective, but also comes fully equipped with a corresponding formal machinery. It thus certainly deserves the attention not only of philosophers, but also of mathematical logicians. Indeed, regarding claims 3, 4, and 5, above, Brandom argues that

> [...] holism *within* each level of constructional complexity is entirely compatible with recursiveness *between* levels. [...] The system I am describing allows us to *prove* it. (In this context, proof is the word made flesh.) The semantic values of all the logically compound sentences are computable entirely from the values of *less complex* sentences. (Brandom, 2008, p. 135)

The main purpose of this contribution is to point out an important gap in Brandom's argument that mainly concerns the intended meaning of negated sentences as 'minimal Aristotelean contraries'. If our analysis is correct, claim 4 and consequently also claim 5 remains unsubstantiated.

The rest of the paper is organized as follows. We start with a short review of Brandom's axioms for incoherence and incompatibility Section 2). This is followed by a discussion of Brandom's concepts of holism and of recursive projectibility in Section 3. Section 4 addresses what I perceive as the major problem of incompatibility semantics. Throughout Sections 3 and 4 we provide direct citations of (Brandom, 2008) to enable also readers that are not familiar with Brandom's original presentation to judge the adequateness and fairness of our criticism. We conclude in Section 6 with the suggestion to consider logical dialogue games as a pragmatist, analytic, and inferentialist alternative to incompatibility semantics.

2 Incompatibility semantics in a nutshell

Incompatibility semantics is defined for a classical propositional language enriched by a standard modal operator $\Box$.[2] A language L_P is a set of formulas that contains a (finite or infinite) set P of propositional variables and is closed under subformulas: $\neg F \in L_P$ implies $F \in L_P$, $F \wedge G \in L_P$ implies $F, G \in L_P$, and $\Box F \in L_P$ implies $F \in L_P$. Whenever no particular set of variables P is referred to, we will suppress the subscript. For sake of conciseness, let us call any subset of L a *theory*. (We emphasize that no closure under logical consequence is implied: 'theory' is just our abbreviation for 'element of the powerset of L'.) Brandom presents his framework in axiomatic form. The basic semantic notion is *'incoherence'*: theories are classified as either incoherent or else coherent. The set of all incoherent theories is called *incoherence frame* Inc. If we add further formulas to an incoherent theory it remains incoherent. In other words, Inc has to satisfy the following monotonicity condition:

Axiom (Persistence): $X \in \text{Inc}$ and $X \subseteq Y$ implies $Y \in \text{Inc}$.

Given Persistence, a frame induces an *incompatibility function* I mapping theories into sets of theories by stipulating

Axiom (Partition): $X \cup Y \in \text{Inc}$ iff $X \in I(Y)$.

Note that, instead of starting with an incoherence frame, one could start by specifying for any theory X the set of theories $I(X)$ that are incompatible with X. Obviously X is incoherent iff X is incompatible with itself ($X \in I(X)$). Thus the 'Partition Axiom' just amounts to an alternative presentation of incoherence.[3] The function I supports a concise definition of corresponding notion of entailment: a theory X *(incompatibility) entails* a finite set of formulas Y if everything that is incompatible with all formulas in Y is also incompatible with X.

Definition (Entailment): $X \models_{\text{Inc}} Y$ iff $\bigcap_{F \in Y} I(\{F\}) \subseteq I(X)$.

---

[2] We use the signs $\neg$, $\wedge$, and $\Box$, instead of Brandom's N, K, and L, respectively. Moreover we stick to the usual infix notation for conjunction.

[3] One might argue that the presented axioms are incomplete in a rather trivial sense: certainly the empty set of formulas is to be regarded as coherent if one wants to avoid interpretations in which *every* theory is incoherent and every formula is incompatible with itself. However Brandom explicitly admits also the degenerate case where $\{\} \in \text{Inc}$.

This more general form of an entailment relation, where a finite set of formulas, not just a single formula, appears on the right hand side, is just like in Gentzen's sequent calculus **LK** for classical logic; i.e., the right hand side is to be interpreted disjunctively, while the formulas on the left hand side are to be interpreted conjunctively. As usual, we will write $F_1, \ldots, F_n \models_{\text{Inc}} G_1, \ldots, G_m$ instead of $\{F_1, \ldots, F_n\} \models_{\text{Inc}} \{G_1, \ldots, G_m\}$.

Brandom specifies the semantics of logical connectives by the following axioms:

Axiom (Negation): $X \cup \{\neg F\} \in \text{Inc}$ iff $X \models_{\text{Inc}} F$.

Axiom (Conjunction): $X \cup \{F \wedge G\} \in \text{Inc}$ iff $X \cup \{F, G\} \in \text{Inc}$.

Axiom (Necessity): $X \cup \{\Box F\} \in \text{Inc}$ iff $X \in \text{Inc}$ or $\exists Y \notin I(X)$: $Y \not\models_{\text{Inc}} F$.

Disjunction ($\vee$) and implication ($\rightarrow$) are defined from $\neg$ and $\wedge$ as in classical logic.

Incompatibility semantics relates to traditional Tarski-style semantics as summarized in the following theorem, where $\models_{S5}$ is the standard entailment relation (defined over Kripke models) for the modal logic $S5$.

Theorem 1. *For all theories X and formulas F: $X \models_{S5} F$ iff $X \models_{\text{Inc}} F$ for all incompatibility frames* Inc.

Incidentally, the modal component of the language will not really concern us much here. The problem that we want to point out and analyze below arises already for the classical propositional connectives.

3 Holism and recursive projectibility

As already mentioned in the introduction, Brandom insists on the *holism* and *non-compositionality* implied by the way he sets up his semantic framework. Indeed, the axioms for negation and for necessity take a non-predicative form. They refer to entailment and therefore to incompatibilities between theories that are not mentioned on the left hand sides of the equivalences. At least *prima facie*, it is not clear whether the axioms for $\neg F$ and for $\Box F$ are equivalent to conditions that only involve F and the context X, since incompatibility entailment implicitly refers to *all* theories over the given language and not just to those consisting of formulas with lower logical complexity than F and the members of X. (Only the axiom for conjunction amounts to a recursively checkable condition in the obvious way.) Indeed, Brandom states:

> Crucial to the compositionality of meaning is that the semantic values of logically complex sentences be reducible to the semantic values of their constituents. In the framework of incompatibility logic, however, meaning is *holistic*, and so [...] reduction cannot proceed sentence by sentence. (Brandom, 2008, p. 147)

In light of the fact that incompatibility entailment (for the language without $\Box$) coincides with the classical entailment relation as specified by Gentzen's well known introduction rules of **LK** in a manner that only refers to the immediate subformulas of introduced formulas, this claim is somewhat problematic. After all, well known work in *proof theoretic semantics* — see, e.g., (Kahle & Schroeder-Heister, 2006) — where the logical rules of an appropriate (cut-free) sequent calculus or, equivalently, of a normalizing natural deduction system, are regarded as specifications of the *meaning* of logical connectives, can hardly be classified as *holistic* semantics. Nevertheless, it may be granted that there is a sense in which the axioms presented in Section 2 can be read as a 'holistic specification' of the semantics of the logical language L.

More importantly, the above quoted passage continues as follows:

> What we want instead is to show how the frame for a language with logically complex sentences can be reduced to the frame for a syntactically less complex fragment of the language. (Brandom, 2008, p. 147)

Risking the charge of pettiness, in order to argue that my analysis below is on target, I have to point out that in the only relevant interpretation of this central claim, the first occurrence of 'the frame' in the quoted sentence must be read as 'any frame', whereas the second occurrence of 'the frame' must be read as 'some frame'. This should be uncontroversial: after all, employing incompatibility semantics to interpret a concrete formal language L amounts to the specification of a concrete frame for L. Certainly *all* frames for L, i.e., all structures where Inc and I satisfy the above axioms, are to be taken as candidates for interpretation.

In any case, also the passage in (Brandom, 2008, p. 135), cited in the introduction of the present paper, makes clear: recursive projectibility can only be established by showing that the problem of computing the semantic value of a logically compound formula $F \in L$ in any given frame Inc for L can be reduced to computing the semantic values of less complex formulas (not necessarily subformulas of F) in frames that arise from Inc by restriction to the relevant sublanguages of L. Brandom thinks that the key to showing this is the notion of *inferentially conservative extensions* of frames, to be explored in the next section.

4   Problems with inferential conservativeness

Brandom defines a language L' to be a *proper extension* of L if $L \subseteq L'$ and all atomic formulas of L' are already in L. Strictly speaking, Brandom's formulation of the axioms reviewed in Section 2 implies that *all* formulas over a set P of atomic formulas (propositional variables) that can be recursively defined using the connectives $\neg$, $\wedge$, and $\Box$ have to be present in the language over which the frame is defined. (E.g., according to the axiom for conjunction, $\{F, G\} \in$ Inc implies $F \wedge G \in$ Inc and thus the underlying language L has to be closed under conjunction.) Consequently, there were no relevant proper extensions of L, but L itself. However, from his somewhat idiosyncratic definition of a language (reviewed Section 2) and from the comments following the definition of 'proper extension' (Brandom, 2008, p. 147), it is clear that Brandom wants to refer to sets of formulas that are closed under subformulas, but that are not necessarily closed under connecting already present formulas by the logical connectives. The simplest way to save Brandom's intentions is to assume that the axioms for negation, conjunction, and necessity are augmented by the conditions 'if $\neg F \in L$', 'if $F \wedge G \in L$', and 'if $\Box F \in L$', respectively. This assumption allows us to follow Brandom in speaking of languages and corresponding frames that need not be closed under applying logical connectives. A frame Inc$'$ for a language L', that properly extends L, is called *inferentially conservative with respect to* a frame Inc for L if for all $X, Y \subseteq L$ we have $X \models_{\text{Inc}} Y$ iff $X \models_{\text{Inc}'} Y$.

Brandom realizes that there are problems with the desired uniqueness of inferentially conservative extensions of frames in case of infinite theories. Therefore he defines the notion of a 'determined frame' as follows:

> Let L' be a proper extension of L and Inc be a frame for L. The *frame for L' determined by* Inc is the smallest frame for L' that is IC [inferentially conservative] with respect to Inc. (Brandom, 2008, p. 148)

At first sight the use of the definite article, implying uniqueness, seems problematic. However 'smallest' here does *not* mean 'not contained in any other frame with the relevant property', but rather, as clarified by a preceding remark,

> [...] 'smallest' has the sense of, contained in every other frame for L' that is IC with respect to Inc. (Brandom, 2008, p. 148)

This allows Brandom to announce:

> We now show that the determined frame exists. (If it does exist, it is immediate from the definition that it is unique.) (Brandom, 2008, p. 148)

It is indeed not difficult to establish the existence of frames that are inferentially conservative over a given frame for a proper sublanguage. Uniqueness, as indicated, is less straightforward. (Brandom credits his research assistant Alp Aker for crucial technical contributions.)

Theorem 2. *For every proper extension L' of a language L and for every frame* Inc *for L there exists a frame* Inc'_d *for L' that is determined by* Inc.

The problem with Theorem 2 is not that it were wrong but that it does not match the purpose for which it was motivated. Remember from Section 3 that Brandom claims that although incompatibility semantics is holistic and non-compositional, it nevertheless enjoys recursive projectibility. To support this claim formally the following assertion, that is a kind of *inverse* to Theorem 2, has to be considered:

(Strong projectibility): *For every proper extension L' of a language L, every frame* Inc' *for L' is determined by some frame* Inc *over L.*

Arguably, it might suffice to establish the following slightly weaker version:

(Weak projectibility): *For every frame Inc' over L'_P there is some frame Inc over $L_P = P$, such that Inc the frame for L' is determined by Inc.*

However, it is not difficult to show that both forms of projectibility fail. (A concrete counter example will be specified below.) In contrast to Brandom's claims — e.g., in the cited passages in (Brandom, 2008, p. 135 and p. 147) — the semantic values, i.e., the incompatibilities as specified by a given frame, of logically compound formulas are *not* computable from the values of less complex formulas, in general. In fact, already a simple counting argument should make clear that there is *no deterministic way at all* in which a given frame for L' can be reduced to a frame over a proper sublanguage L: in general, if L' is a proper extension of L, there are strictly more different frames for L' than for L. Thus there is no surjective function from the set of frames for L to the set of frames for L'. To illustrate the problem with Brandom's suggestion of using inferential conservativeness to establish recursive projectibility consider the following simple example.

Let the language $L_P = P$ consist of just two atomic sentences (propositional variables), say, $P = \{a, b\}$. If we exclude the possibility that the empty set is incoherent and assume that neither a nor b is self-incompatible, just two possible frames, i.e., two different sets of incoherent subsets of L_P remain:

- $\text{Inc}_1 = \{\{a, b\}\}$, meaning that a and b are incompatible,

- $\text{Inc}_2 = \{\}$, meaning that a and b are not incompatible.

By the definition of (incompatibility) entailment, we have $a \not\models_{\text{Inc}_1} b$ and $b \not\models_{\text{Inc}_1} a$, but $a \models_{\text{Inc}_2} b$ and $b \models_{\text{Inc}_2} a$. Moreover, note that for *all* frames Inc over L_P and all $F, G \in L_P$ we have: $F \models_{\text{Inc}} G$ iff $\{F, G\} \notin \text{Inc}$. The fact that compatibility (i.e. non-incompatibility) between formulas implies entailment already hints at a problem. However, at this point, one might still be satisfied with the remark that the strange coincidence of compatibility and entailment is due to the absence of negation from the language. Let us therefore consider the proper extension $L'_P = \{a, b, \neg a, \neg b\}$ of L_P. It is easy to check that the following two frames Inc'_1 and Inc'_2 over L'_P are inferentially conservative with respect to Inc_1 and Inc_2, respectively:

Incompatibility Semantics 89

- Inc'_1 consists of those theories over L'_P that contain $\{a,b\}$, or $\{\neg a, a\}$, or $\{\neg b, b\}$. (Note that, because of Persistence, a frame cannot just consist in the three exhibited two-element sets. We always have to close of with respect to supersets to obtain a frame.)

- Inc'_2 consists of those theories over L'_P that contain $\{a, \neg b\}$, or $\{\neg a, b\}$, or $\{\neg a, a\}$, or $\{\neg b, b\}$.

Since Inc'_1 is inferentially conservative over Inc_1, we still have $a \not\models_{Inc'_1} b$ and $b \not\models_{Inc'_1} a$. More generally, we obtain $F \not\models_{Inc'_1} G$ for all $F, G \in L'_P$, where $F \neq G$. In contrast, for Inc'_2, we obtain $a \models_{Inc'_2} b$ and $b \models_{Inc'_2} a$, as required for the inferential conservativeness of Inc'_2 with respect to Inc_2. Moreover we have $\neg a \models_{Inc'_2} \neg b$ as well as $\neg b \models_{Inc'_2} \neg a$ by the form of Inc'_2 and the axioms. Note that in all frames each formula is incompatible with its negation (assuming, of course, that it is in the language). Moreover all frames that are inferentially conservative over Inc_1 have to contain $\{a, b\}$, since $\{a, b\} \in Inc$ is equivalent to $\{a, b\} \models_{Inc} \{\}$ for all frames Inc. Therefore Inc'_1 is the smallest (and in fact the only) frame for L'_P that is inferentially conservative with respect to Inc_1. Thus it is *determined* by Inc_1, according to Brandom's definition. The case for Inc'_2 is similar: it is the only frame for L'_P that is inferentially conservative over Inc_2. Consequently Inc'_2 is *determined* by Inc_2.

For our argument it is important to recognize that the above examples of determined frames leave many possible frames for L'_P as *not determined by any frame* for L_P. In particular, it is straightforward to check that in the only frame Inc for L_P where $a \models_{Inc} b$ and $b \not\models_{Inc} a$ the atomic formula a is self-incompatible, which implies that $a \models_{Inc} \{\}$. But it is not difficult to specify a frame Inc' for L'_P where $a \models_{Inc'} b$ and $b \not\models_{Inc'} a$ as well as $a \not\models_{Inc} \{\}$: let Inc' consist of just those theories that contain $\{a, \neg b\}$ or $\{a, \neg a\}$ or $\{b, \neg b\}$. Inc' is not inferentially conservative over *any* frame for L_P and thus cannot be recursively projected or in any other systematic way reduced to any frame over L_P. This provides the concrete counter example to (weak and strong) projectibility, announced earlier. As already indicated above, the central fact that not all frames over languages with negation can be reduced to frames without negation can also be established by checking that there are just 5 different frames for L_P, while there exist more than twice as many different frames for L'_P, each inducing a different entailment relation.

5 Consequences of the non-reducibility of frames

As we have seen in the last section, recursive projectibility, i.e. the claim that incompatibilities between logically complex formulas are determined by the incompatibilities between formulas that are less complex, cannot be maintained. Independently of the contents of Theorem 2, the presented examples show that there are frames over simple, finite languages with negation for which the incompatibilities are not determined by any frame over a language without negation. In other words, even the knowledge of the semantic status of *all* formulas without negation does not suffice to determine the semantic status of formulas with negation. Coming back to Brandom's five claims about incompatibility semantics, formulated in the Introduction of this paper, this means that claim 4 (recursive projectibility) cannot be maintained. Since claim 5, namely that the properties of incompatibility semantics allow to refute the assertion that holistic semantics cannot account for the learnability of a language, depends on the validity of claim 4, it remains unsubstantiated as well.

We emphasize that the outlined problem arises already for non-modal languages. Moreover, the axiom for conjunction stipulates a direct reduction of the status of a formula $F \wedge G$ to that of F and G, respectively, and thus does not contribute to the 'holism' or to the intensional character of incompatibility semantics. Brandom insists in defining disjunction and (material) implication in terms of negation and conjunction, just like in classical logic: $F \vee G =_{\text{df.}} \neg(\neg F \wedge \neg G)$ and $F \rightarrow G =_{\text{df.}} \neg(F \wedge \neg G)$. Therefore, as far as non-modal languages are concerned, the problem rests with the suggested semantics of negation. However, in light of claim 2 (intentionality), claim 3 (holism and non-compositionality), but also claim 5 (learnability), it might be deemed odd that Brandom does not consider more direct, alternative routes to provide meaning for disjunction and, in particular, implication. Of course, if the only aim were to characterize ordinary classical logic or the simplest modal logic, $S5$, in terms of incoherence/incompatibility, then restricting attention to negation and conjunction is an obvious move. However, why should one insist on classical logic or on $S5$ in the wider context of inferentialism and analytic pragmatism? Indeed, some of Brandom's remarks, in particular in his reply to an attempt to bring incompatibility semantics for modal logic closer to Kripke semantics (Göcke, Pleitz, & Wulfen, 2008),[4] indicate that Brandom were happy to go beyond just $S5$. Concerning the non-modal fragment of the language there are corresponding remarks on intuitionism in (Brandom, 2008). But remember that, while $\neg F$ abbreviates $F \to \bot$, falsum ($\bot$), disjunction, conjunction, and implication are not inter-definable in intuitionistic logic. In any case, I cannot imagine that Brandom really wants to claim that, e.g., the stipulation that $F \vee G$ abbreviates $\neg(\neg F \wedge \neg G)$ is all one must or should say about the learnability of disjunctive sentences, granted that we know how to learn the meanings of negation and conjunction. Consequently, the case for claim 5 would remain incomplete, even if recursive projectibility could be established.

There seems to be an easy way out of the quagmire of frames for negation-free languages: just insist that all *literals*, i.e., all atomic formulas and all their negations are always present in a language. However this suggestion does not fit well with Brandom's explicit intention to treat classical propositional connectives in an intensional manner (cf. claim 2), on a par with the modal operator, and without involving the notions of truth or falsity. Given the 'recursive' axiom for conjunction and the definition of disjunctions and implications as abbreviations of negated conjunctions, we are left with just a rather trivial notational variant of ordinary Tarskian truth functional semantics for classical logic. To see this, we identify classical interpretations, i.e. assignments of true and false to propositional variables $\{p_1, p_2, \ldots\}$, with maximally coherent subsets of literals $\{p_1, \neg p_1, p_2, \neg_2, \ldots\}$. More formally, any classical truth value assignment $v : P \mapsto \{true, false\}$ induces the set $\Phi(v) = \{p \mid v(p) = true\} \cup \{\neg p \mid v(p) = false\}$. Clearly Φ amounts to a bijection between truth value assignments and maximally coherent sets of literals, since every set that contains both p and $\neg p$ for some propositional variable $p \in P$ is incoherent,

---

[4]This is not the place to analyze the interesting ideas of Göcke, Pleitz, and von Wulfen. However, let me emphasize that some of their suggestions are clearly mistaken for precisely the same reason that invalidates some of Brandom's claims: worlds in a Kripke structure cannot be identified with sets of atomic sentences that are maximally coherent with respect to an incompatibility frame, since those sets, even jointly, do not determine the semantic status of negated sentences and of modal sentences.

and since every coherent set of literals that contains neither p nor $\neg p$ can be extended to another coherent set by adding either p or $\neg p$. In other words: there remains only a superficial difference between talk about (Tarskian) interpretations satisfying a formula F and talk about maximally coherent subsets of literals that incompatibility-entail F, respectively. This can hardly be accepted as a way to save incompatibility semantics as intended.

6 Hints on dialogue game semantics

After this largely negative assessment of some central aspects of incompatibility semantics, we want to end at a more positive note by briefly suggesting that Brandom's fascinating project of providing a formal semantics that fits well into the wider frame of inferentialism and analytic pragmatism could well be based on quite different notions. In fact, I can think of many different ways of building formal semantics on pragmatist and inferentialist grounds. At least one concept of this kind is ready to be picked up from the literature: dialogue game semantics.

Already in the late 1950s Paul Lorenzen suggested to specify the meaning of logical connectives by reference to what we are obliged and entitled to do in confrontational dialogues involving logically complex assertions (Lorenzen, 1960). More precisely, Lorenzen suggested a strategic game that starts with the assertion of a sentence (formula) F by a proponent P to be challenged by an opponent O and proceeds by systematic attack and defense moves referring to relevant subformulas of F. The rules guiding this game take the following form:

Implication rule: If P asserts $F \to G$ then O is entitled to attack by asserting F in reply, which in turn obliges P to assert G as well.

Conjunction rule: If P asserts $F \wedge G$ then O is entitled to attack by obliging P to assert F, likewise O is entitled to attack by obliging P to assert G.

Disjunction rule: If P asserts $F \vee G$ then O is entitled to attack by obliging P to either assert F or to assert G, where the choice is up to P.

The rules have been formulated just for P as defender as O as attacker, but the implication rule entails that the roles may switch. Accordingly, analogous rules also hold for inverted roles, i.e. for P challenging assertions of O. $\neg F$ is defined as $F \to \bot$, where $\bot$ is a formula that can never be defended successfully; i.e. whoever asserts $\bot$ looses the dialogue game. Otherwise P wins the game if O attacks a sentence that O herself has already asserted previously. If furthermore certain structural rules (Lorenzen's *Rahmenregeln*), regulating the succession of attacks and defenses, are imposed, one can show that P has a winning strategy for exactly those initial assertions that are intuitionistically valid, see (Felscher, 1985). Already Lorenzen and his collaborators considered different versions of the game — for further references see (Felscher, 1986), (Krabbe, 1985) — and it soon became clear that not only intuitionistic logic, but rather a large variety of logics, including classical logic and modal logics, can be characterized in a similar manner. Moreover, systematic correspondences between logical dialogue games and sequent as well as hypersequent based proof theory have been explored, e.g., in (Krabbe, 1985), (Fermüller, 2003), (Fermüller & Metcalfe, 2009). Those correspondences incidentally relate dialogue games also to the programme of proof theoretic semantics, mentioned briefly in Section 3. In the 1970s a particularly interesting variant of Lorenzen's ideas has been developed by Robin Giles in an attempt to provide 'tangible meaning' to logical connectives and atomic assertions as they arise in reasoning within theories of physics (Giles, 1974), (Giles, 1977). Giles combines a Lorenzen style dialogue game with a betting scheme on the results of elementary experiments associated with atomic sentences that may show dispersion; i.e. the same experiment may yield a different result upon repetition — only a subjective success probability is stipulated. Giles showed that a strategy, that guarantees that no money is lost in average when arguing and betting accordingly, exists if and only if the originally asserted sentence corresponds to a formula that is valid in infinitely value Łukasiewicz logic. This result has later been generalized to further so-called t-norm based fuzzy logics and, again, connected to proof theoretical investigations (Ciabattoni, Fermüller, & Metcalfe, 2005),(Fermüller & Metcalfe, 2009), (Fermüller, 2009).

Admittedly, these short and vastly incomplete hints at dialogical approaches to logic cannot replace a serious investigation that seeks to clarify whether, why, and how the largely *implicit* reference to normative, pragmatic, and inferentialist concepts in dialogue games can be made *explicit*. I can only hope that my critical remarks on incompatibility semantics don't deter any friend of Brandom's general approach to semantics and to reasoning — among whom I certainly count myself — from taking up this suggestion.

Christian G. Fermüller
Theory and Logic Group, TU Wien
Favoritenstr. 9-11, A-1090 Vienna, Austria
chrisf@logic.at
http://www.logic.at/staff/chrisf

References

Brandom, R. (2008). *Between saying and doing. Towards an analytic pragmatism.* Oxford University Press.

Ciabattoni, A., Fermüller, C., & Metcalfe, G. (2005). Uniform Rules and Dialogue Games for Fuzzy Logics. In *Proceedings of LPAR 2004* (Vol. 3452, pp. 496–510). Springer.

Felscher, W. (1985). Dialogues, strategies, and pntuitionistic provability. *Annals of Pure and Applied Logic, 28*, 217–254.

Felscher, W. (1986). Dialogues as foundation for intuitionistic logic. In D. Gabbay & F. Günther (Eds.), *Handbook of philosophical logic* (Vol. III, pp. 341–372). Reidel.

Fermüller, C. (2003). Parallel dialogue games and hypersequents for intermediate logics. In M. C. Mayer & F. Pirri (Eds.), *Proceedings of TABLEAUX 2003* (Vol. 2796, pp. 48–64). Springer.

Fermüller, C. (2009). Revisiting Giles — connecting bets, dialogue games, and fuzzy logics. In O. Majer, A. Pietarinen, & T. Tulenheimo (Eds.), *Games: Unifying logic, language, and philosophy* (pp. 209–227). Springer.

Fermüller, C., & Metcalfe, G. (2009). Giles's game and the proof theory of Łukasiewicz logic. *Studia Logica*(92), 27–61.

Giles, R. (1974). A non-classical logic for physics. *Studia Logica, 4*(33), 399–417.

Giles, R. (1977). A non-classical logic for physics. In R. Wojcicki & G. Malinkowski (Eds.), *Selected papers on Lukasiewicz sentential calculi* (pp. 13–51). Polish Academy of Sciences.

Göcke, B., Pleitz, M., & Wulfen, H. von. (2008). How to Kripke Brandom's notion of necessity. In B. Prien & D. Schweikard (Eds.), *Robert Brandom, Analytic Pragmatist* (pp. 135–147). Frankfurt: Ontos Verlag.

Kahle, R., & Schroeder-Heister, P. (Eds.). (2006). *Special issue on proof theoretic semantics* (Vol. 148). Synthese.

Krabbe, E. (1985). Formal systems of dialogue rules. *Synthese*(63), 295-328.

Lorenzen, P. (1960). Logik und Agon. In *Atti congr. internaz. di filosofia* (pp. 187–194). Sansoni.

From $(AB)a$ infer A^*a

Bjørn Jespersen*

1 Introduction

This paper makes a contribution to the logic of property modification by introducing a new rule that is required to validate an existing rule of property modification. The new rule also validates various inferences that I argue ought to go through.

Let a property be a function from possible worlds to functions from times to sets of individuals, to capture both the modal and temporal variability of the extensions of empirical properties. Let a property modifier be a function from properties to properties.[1] Where A is a modifier, B, A^* properties and a an individual, the four rules that are standardly taken to exhaust property modification are (in rudimentary notation):[2]

Subsective. From $(AB)a$ infer Ba.
Example. If a is a skilful surgeon then a is a surgeon.

---

*I am indebted to Roberto Ciuni, Marie Duží, Lloyd Humberstone, and Greg Restall for valuable comments. Versions of this paper were read at LOGICA2009, Hejnice, SOPHA 2009, Geneva, and School of Philosophy, Anthropology and Social Inquiry, University of Melbourne, 27 November 2009. This paper builds in part on material also appearing in (Duží, Jespersen, & Materna, 2010, §4.4).
The research reported herein has been pursued within the project GA CR 401/10/0792 *Temporal aspects of knowledge and information*.

[1]That is, a modifier operates on *intensional* entities, but is not itself one, being a function not indexed to possible worlds that obtains between intensions. The discussion in Heim and Kratzer (Heim & Kratzer, 1998, p. 72) of 'former', as it occurs in 'is a former lover of Bill's' and 'is a former tenant of 13 Green St' (to use their example), goes to show that modifiers should not be functions from *sets* to *sets*. In their example, it so happens that the class of Bill's former lovers and the class of former tenants of 13 Green St are the same class. But the properties of being a former lover of Bill's and being a former tenant of 13 Green St ought to have the same satisfaction class only contingently, whereas the typing $\langle\langle e,t\rangle,\langle e,t\rangle\rangle$ predicts that they must have the same satisfaction class of necessity. Heim and Kratzer point out that since 'former' has a temporal, or dynamic, dimension (since a former F no longer is, though used to be, an F), this modifier demands an intensional semantics. The authors, however, as they themselves acknowledge, leave the topic of intensional semantics underdeveloped.

[2]See, e.g., (Clark, 1970), (Kamp & Partee, 1995) and (Partee, 2001).

Intersective. From $(AB)a$ infer $A^*a \wedge Ba$.
Example. If a is a happy child then a is happy and a is a child.
Modal. From $(AB)a$ infer $Ba \vee \neg Ba$.
Example. If a is an alleged assassin then a is an assassin or a is not an assassin.
Privative. From $(AB)a$ infer $\neg Ba$.
Example. If a is a fake banknote then a is not a banknote.

The fifth rule, which I suggest, is the rule of pseudo-detached modification:

Pseudo-detached. From $(AB)a$ infer A^*a.

This rule imitates, as it were, the effect of detaching A from $(AB)a$. In fact, a sloppy statement of intersection — "From $(AB)a$ infer $Aa \wedge Ba$" — would suggest that A could actually be detached from (AB) and predicated of a. Proper detachment is invalid, however, since A is a modifier and so applicable to B but inapplicable to a. Hence, attempting to predicate A of a would be tantamount to the type-theoretic error of juxtaposing a property modifier and an individual. But if A is pseudo-detached, this type-theoretic error is replaced by the type-theoretically proper juxtaposition of a property and an individual. (Below, though, I shall qualify this claim, in that what gets applied to the individual is an extensionalized property, which is a characteristic function from individuals to truth-values.)

An example to fix ideas. If the premise is that a is a black cat and if the adjective 'black' denotes a modifier and the noun phrase 'cat' denotes a property, then the conclusion is that a is black. The logical form of the conclusion is that there is a property p such that a is a (black p), where (black p) is the property resulting from applying the modifier *black* to the value of the property variable p. Abbreviate the predicate 'is a (black p)' as 'black*'. Then, while the adjective 'black' denotes a modifier, the adjective 'black*' denotes a property. This is to say that 'black' does not change its denotation from a modifier to a property. It only seems as though it would do so because in the surface grammar the same adjective occurs twice over when generating the sentence, "A black cat is black".

Note that only subsective, privative and pseudo-detached modification need to be introduced as primitive operations, while intersective and modal modification lend themselves to being defined in terms of those three.³ Thus, modal modification can be defined in terms of subsective or privative modification, together with *or* introduction.⁴ And intersective modification can be defined in terms of subsective and pseudo-detached modification, together with *and* introduction.

Let me emphasize that, and why, pseudo-detachment is a necessary prerequisite for intersective modification. Without this rule, or any other rule to the same effect, "a is a happy child" could not be validly factored out into "a is happy" and "a is a child". The fact that the first occurrence of 'happy' denotes a modifier, and the second a property, is not always appreciated. For instance, the influential Heim and Kratzer (Heim & Kratzer, 1998) introduces a rule of intersective modification (dubbed 'PM', short for 'Predicate Modification') that is not accompanied by a type-shifting rule for replacing a modifier by a property (or a predicate, if the mode is formal and not material). Thus, PM allows that if x is a city in Texas then x is a city and x is in Texas (Heim & Kratzer, 1998, p. 65), without laying down how 'is a city in Texas', 'in Texas' denoting a modifier, may be factored out into 'is a city' and 'is in Texas', both denoting a property.⁵ Formally, relative to Heim and Kratzer's framework, such a rule would have had to sustain swapping a term of the modifier type $\langle\langle e,t\rangle,\langle e,t\rangle\rangle$ for a term of the same syntactic shape of the property type $\langle e,t\rangle$. At the same time, though, Heim and Kratzer are fully alert to the fact that some occurrences of, e.g., the adjective 'small' assign the type $\langle e,t\rangle$ to it while other occurrences assign the type $\langle\langle e,t\rangle,\langle e,t\rangle\rangle$. But Heim and

---

³See (Jespersen & Primiero, forthcoming) for two variants of a procedural semantics (one based on Transparent Intensional Logic, the other on Martin-Löf's Constructive Type Theory) for privative modifiers.

⁴Logically speaking, modal modification is trivial, since its conclusion is no more than a classical tautology, which is also satisfied by the other kinds of modifier. For instance, if a is a skilful surgeon then it follows that a is a surgeon or that a is not a surgeon; it is just that the first disjunct always holds and the second one never does. What is unique about modal modification is that there are cases where the respective first disjunct holds and other cases where the second disjunct holds. As for 'modal modifier', this is the term used by, e.g., Partee; see (Partee, 2001, p. 7). Such modifiers are also known as 'intensional' (Cresswell, 1978) and 'equivocating' (Coulson & Fauconnier, 1999). See also (Rotstein & Winter, 2004, p. 276, n. 14).

⁵To circumvent the locative phrase 'in Texas', we might opt for 'Texan' instead to generate the predicate 'is a Texan city'. Thus, from "x is a Texan city" we get "x is Texan", i.e., "x is a (Texan something)", and "x is a city".

Kratzer's discussion of 'small' as it occurs in, e.g., "Jumbo is small" (Heim & Kratzer, 1998, p. 70, entry 18) remains inconclusive.[6]

2 The rule of pseudo-detachment

Let $[AB]$ be the property resulting from applying A to B, a pair of square brackets representing functional application, and let $[[[AB]t]w]$, abbreviated '$[AB]_{wt}$', be the result of applying the property $[AB]$ to the values of the world and time variables w, t to obtain a set, in the form of a characteristic function, applicable to a.[7] Further, let $=$ be the identity relation between properties, and let p range over properties, x over individuals. Then the *proof* of the rule "From $(AB)a$ infer A^*a" is this:

1. $[[AB]_{wt}a]$ assumption
2. $\exists p[[Ap]_{wt}a]$ 1, EG
3. $[\lambda x \exists p[[Ap]_{wt}x]a]$ 2, λ-expansion
4. $[\lambda w' \lambda t'[\lambda x \exists p[[Ap]_{w't'}x]]_{wt}a]$ 3, λ-expansion
5. $A^* = \lambda w' \lambda t'[\lambda x \exists p[[Ap]_{w't'}x]]$ definition
6. $[A^*_{wt}a]$ 4,5, Leibniz's Law

Any valuation of the free occurrences of the variables w, t that makes the first premise true will also make the second, third and fourth steps true. The fifth premise is introduced as valid by definition. Hence, any valuation of w, t that makes the first premise true will, together with step five, make the conclusion true.[8] Therefore, the following *argument* is valid:

$$\frac{\lambda w \lambda t[[AB]_{wt}a];\ A^* = \lambda w' \lambda t'[\lambda x \exists p[Ap]_{w't'}x]]}{\lambda w \lambda t[A^*_{wt}a]}$$

---

[6]It may, or may not, follow from my stance on pseudo-detachment that a sentence like "Jumbo is small" is elliptical for "Jumbo is a (small p)", with a suppressed reference, in the form of a free variable, to some property or other. (Note that "$\exists p$ (Jumbo is a (small p))" is trivial, while "Jumbo is small" is not.) But, for the record, I would be happy to embrace such an analysis of "Jumbo is small". It will fall to the pragmatic or linguistic context that "Jumbo is small" is embedded in to provide a specific value for p. Whenever the parties to a discourse are in on what property is being suppressed, there is no pragmatic need to make it explicit. But it always remains legitimate to ask, "Jumbo is a small *what*?", and the speaker must cite a specific property to satisfy the hearer. (This addresses a question that came up during discussion time at LOGICA.)

[7]See (Jespersen, 2008) for details on extensionalization of properties in order to predicate them of individuals.

[8]For λ-binding w, t, see (Tichý, 1971) and (Jespersen, 2005).

As the above proof and argument both make clear, the rule does not add expressive power to the (fragment of the) typed intensional lambda calculus in which it is stated and validated. Rather the rule deploys only standard terms and techniques. But the merit of the rule is, at the very least, that it spells out something that should have been spelt out long ago, in order to validate the first conjunct of intersective modification, *in casu* "a is happy" and "x is in Texas". Otherwise we will be stuck for an answer to the question of why it is logically valid to infer from a being a happy child that a is happy and from x being a city in Texas that x is in Texas.

It is crucial not to confuse the rule of pseudo-detachment with the following fallacy, which Geach rightly objects to:[9]

> a is a big flea, so a is a flea and a is big; b is a small elephant, so b is an elephant and b is small; so a is a big animal and b is a small animal. (Geach, 1956, p. 33)

Pseudo-detachment licenses no such conclusion, however. Geach's illicit move is to steal the property *being an animal* into the conclusion, thereby making a, b commensurate. Both fleas and elephants are animals, to be sure, but a's being big and b's being small follow from a's being a *flea* and b's being an *elephant*, so pseudo-detachment only licenses the following two inferences, where the values of the variables p, q are two distinct properties:

> There is a property p such that a is a (big p); there is a property q such that b is a (small q).

And a (big p) may well be smaller than a (small q), depending on the values assigned to p, q.

This outcome ought to strike us as being pretty much trivial, anyway. If we grant, uncontroversially, that nobody and nothing is absolutely small or absolutely big then everybody and everything is made small by something and made big by something else. Heim and Kratzer seem to be saying something to roughly the same effect:

> We do predict that [(13) "Jumbo is a small elephant"] entails [(14) "Jumbo is a small animal"] *if the context for both is the same.* (Heim & Kratzer, 1998, p. 71. Their emphasis.)

---

[9] See also (Gamut, 1991, § 6.3.11).

They stress that they do not intend their (14) to mean that Jumbo is small for an average-sized animal, the reason being, no doubt, that a small elephant may well dwarf most other average-sized animals and for this reason be too large to qualify as a small animal. Instead Jumbo should be measured only against other elephants, and since the premise says that Jumbo is small for an elephant, and since all elephants are animals, it follows that there is a species whose current norm of average size is such that Jumbo is below par.

In case we neglect the relativization to context (be it a scale, a norm, etc.), we risk reading too much into pseudo-detachment or any other rule having the same effect. In fact, this appears to be what makes Clark balk at drawing the following inferences which I deem obviously valid:

> We cannot ... infer that something is red from the fact that it is a large, red ball. We cannot infer that John is famous from the fact that John is a famous ichthyologist. This may seem an awkward consequence of the account. Nonetheless, it is, I think, the way things should be. What we can infer from the fact that this is a large, red ball is that it is a large ball, and that it is a red ball. We do not want it to follow from the fact that this is a large chigger, that this is large. So, too, I believe, we should not want it to follow that the fact that this is a very red chigger, that this is very red. A very red chigger need not be very red at all. Similarly, we cannot correctly infer that John is famous ... from the fact that John is a famous ichthyologist. A famous ichthyologist may be an unknown in the marketplace. (Clark, 1970, p. 334)

Being famous, though, does not require being famous throughout the known universe. John merely needs to be famous in ichthyologist quarters and not also amongst the fishmongers in the marketplace. Since John is a famous ichthyologist, it follows that there is a property p such that John is a (famous p), which is all pseudo-detachment demands. Likewise, if a is a very red chigger, it follows that there is a property q such that a is a (very red q), which is all pseudo-detachment demands. (Similarly for large chiggers being large.) The relativization to q makes it irrelevant that a very red chigger (i.e., a chigger that is exceptionally red for a chigger) may pale in comparison to, say, the Albanian flag.

3 Pseudo-detachment and higher-order modification

Adverbial modifiers like *very* appear to pose a threat to the validity of pseudo-detachment. Here I show why they do not. Consider this sentence:

"*a* is a very large elephant"

If we insert '*Very*' for '*A*' in premise (2) of the proof above, then something seems to go awry between steps (4') and (5'):

4'. $[\lambda w' \lambda t' [\lambda x \exists p [[Very\ p]_{w't'} x]]_{wt} a]$

5'. $Very^* = \lambda w' \lambda t' [\lambda x \exists p [[Very\ p]_{w't'} x]]$

For there is obviously no such property as $Very^*$. Yet the rule of pseudo-detachment appears to license what has at least the form of a legitimate predication:

6'. $[Very^*_{wt} a]$

While we do understand what it would mean that *a* is large (namely, a large something, i.e. large with respect to some property), it is not possible to make sense of *a* being very (full stop), simply because "*a* is very" is ungrammatical and nonsensical for want of a complement for the adverb 'very'. So clearly we need to get clearer about the logical structure underlying "*a* is a very large elephant".

To anticipate, the root of the fallacy is that 'Very' in (4') denotes a property modifier. Fortunately, there is an alternative, and much more common, construal of 'Very', namely as denoting a higher-order modifier, which modifies property modifiers.[10] So let us analyze these two ways of construing 'Very' in detail. On the first construal, the argument of *Very* is [*Large Elephant*], taking the property of being a large elephant to the property of being a very large elephant: for any world/time pair $\langle w, t \rangle$, given its set of large elephants at $\langle w, t \rangle$, we are given its subset of very large elephants at $\langle w, t \rangle$. The modifier

---

[10]See (Rotstein & Winter, 2004, p. 276). Clark (Clark, 1970, p. 328) speaks of 'mod-mods', i.e. modifiers of modifiers, citing 'powerfully' in 'powerfully compelling performance' as an example of a mod-mod.

Very modifies a property that has in turn been generated by applying a modifier to a property. *Very* is inapplicable to *Elephant*, but does apply to [*Large Elephant*]. The predication of the property of being a very large elephant of a, accordingly, has this logical form:

$$\lambda w \lambda t[[\mathit{Very}[\mathit{Large\ Elephant}]]_{wt}a]$$

By substituting the variable p ranging over properties for [*Large Elephant*], we end up with [*Very p*], which is the one that got us into trouble.

On the second construal of 'Very', the argument of Very' is *Large*, thus assigning a different logical type to Very'; namely, the type of a mapping from property modifier to property modifier. Thus, when applying the modifier modifier Very' to the property modifier *Large*, the result is the property modifier [Very' *Large*] which, when applied to *Elephant*, yields the property [[Very' *Large*] *Elephant*]: for any world/time pair $\langle w, t \rangle$, given its set of elephants at $\langle w, t \rangle$, we are given its subset of very large elephants at $\langle w, t \rangle$. The predication of the property of being a very large elephant of a now has this logical form:

$$\lambda w \lambda t[[[\mathit{Very}'\ \mathit{Large}]\mathit{Elephant}]_{wt}a]$$

For a more abstract comparison, quantify away the modifier *Large* by substituting it by f ranging over property modifiers and also quantify away the property *Elephant* by substituting it by p ranging over properties.[11] Applying this dual quantification to the last two logical forms yields, respectively:

$$\lambda w \lambda t[\exists f[\exists p[[\mathit{Very}[fp]]_{wt}a]]] \quad \text{and} \quad \lambda w \lambda t[\exists f[\exists fp[[[\mathit{Very}'f]p]_{wt}a]]]$$

These last two logical forms schematize, by means of variables, the two different scopes of the modifier denoted by 'Very', according as this modifier modifies properties or modifiers. Whether the construal of the adverb 'Very' be as denoting a property modifier or a modifier modifier, the same individuals are worthy of the predicate 'is a

---

[11] For the record, where f ranges over property modifiers, this inference is valid:
$$\lambda w \lambda t[[[\mathit{Very}'\ \mathit{Large}]\mathit{Elephant}]_{wt}a]$$
$$\lambda w \lambda t[\exists f[[f\ \mathit{Elephant}]_{wt}a]]$$
The conclusion means that the property *Elephant* is modified in some way or other.

very large elephant', whatever set of elephants may be involved, so at the level of possible-world properties (i.e. properties identified up to necessary co-extensionality) there is no difference. There is, however, a subtle, yet important, distinction at the level of conceptualization, or generation, of properties. (Think of conceptualizations of possible-world properties as *hyperproperties*, if you like.) The property of being a very large elephant is either conceptualized as the special sort of *large elephant* that is a very large one (i.e. has largeness, as far as elephants goes, to an extreme degree) or is conceptualized as the special sort of *elephant* that is a very large one.

Due to the difference in the arguments of *Very* and *Very'*, the rule of pseudo-detachment applies immediately to $[[\textit{Very' Large}]\textit{Elephant}]$ but not to $[\textit{Very}[\textit{Large Elephant}]]$. If a has the property $[[\textit{Very' Large}]\textit{Elephant}]$, then it follows from the rule of pseudo-detachment that a is very large, i.e. a very large something. Step (5") is this:

5". $[\textit{Very' Large}]^* = \lambda w' \lambda t' [\lambda x \exists p [[[\textit{Very' Large}]p]_{w't'}x]]$

$[\textit{Very' Large}]^*$ is the property of being very large. So the smoothness of the inference to the conclusion

6". $\lambda w \lambda t [[\textit{Very' Large}]^*_{wt} a]$

recommends going with the construal of 'Very' as denoting a modifier modifier. This makes good philosophical sense, anyway. Just as *Large* modifies the property *Elephant* (such that a is not just any old elephant, but a large one), so it is feasible to modify *Large* itself. Largeness comes in various degrees even with respect to the same property F, since a may be a ((fairly large) F) and b an ((extremely large) F).

The elegance of *Very'* does not in itself disqualify *Very*, of course. What does count against *Very* is that it requires an *ad hoc* restriction for pseudo-detachment to be validly applicable. The restriction is that the complement of *Very* must be a property that is measurable in degrees, like *being an expensive house, being a happy logician, being a successful assassin*, etc. *Very* cannot take as a complement a property like *being a child, being a flute* and other such absolute properties, which do not come in degrees.[12] There are instances of $[\textit{Very } p]$ where this restriction is not going to be heeded, unless the quantificational

---

[12] I am sidestepping the issues of vagueness and fuzziness.

range of p is somewhat artificially restricted to properties measurable in degrees. Such a restriction would serve to include sentences like, "a is a very expensive flute" and exclude ill-formed sentences like, "b is (a) very flute". The need for the restriction seems to suggest that *Very* really just reflects the fact that *Very'* modifies modifiers like *expensive* rather than properties like *being a flute*, which in turn seems to suggest that *Very* is little more than a modifier modifier masquerading as a property modifier. This impression is buttressed by the shallowness of

$$\lambda w \lambda t [\exists p [[\mathit{Very}\ p]_{wt} a]]$$

as a logical analysis of the existential quantification of

"a is a very large elephant"

The single property variable p conceals the fact that *being a large elephant* has been generated by applying a modifier to a property, whereas a deeper analysis would have one variable for the modifier and another for the property to be modified. Thus,

$$\lambda w \lambda t [\exists f [\exists p [[[\mathit{Very'}\ f] p]_{wt} a]]]$$

offers more by way of logical depth by making more logical operations explicit. This recommends the higher-order modifier *Very'* over the first-order modifier *Very* as the more satisfactory analysis of the adverb 'Very'. Besides, we have all the more reason to adopt *Very'*, since it was demonstrated above why the occurrence of a higher-order modifier such as this does not compromise the validity of the rule of pseudo-detachment.

4 Conclusion

Pseudo-detachment is in essence a rule for trading the modifier A for the property A^*. The trick is to quantify the property B away and replace it by the $\exists$-bound variable p. So if Jumbo is a small elephant, then Jumbo is small*; i.e. Jumbo is a small something (rather than small *absoluter*); i.e. there is a property p such that Jumbo is a (small p). And if Jumbo is a very small elephant, then Jumbo is (a very small something). Furthermore, the validity of pseudo-detachment is not predicated on whether the modifier involved be intersective, subsective, modal, or privative. Finally, the interplay between pseudo-detachment and higher-order modification was dealt with, showing how and why a predicate like 'is a very large elephant' does not constitute a counterexample to pseudo-detachment.

Bjørn Jespersen,
Department of Computer Science FEI,
VSB–Technical University of Ostrava,
Czech Republic
bjorn.jespersen@gmail.com

References

Clark, R. (1970). Concerning the logic of predicate modifiers. *Noûs*, *4*, 311–335.

Coulson, S., & Fauconnier, G. (1999). Fake guns and stone lions: conceptual blending and privative adjectives. In B. Fox, J. D., & L. Michaelis (Eds.), *Cognition and function in language*. Palo Alto: CSLI.

Cresswell, M. (1978). Prepositions and points of view. *Linguistics and Philosophy*, *2*, 1–41.

Duží, M., Jespersen, B., & Materna, P. (2010). *Procedural semantics for hyperintensional logic. Foundations and applications of transparent intensional logic*. Berlin: Springer–Verlag.

Gamut, L. (1991). *Logic, language and meaning 2*. Chicago–London: University of Chicago Press.

Geach, P. (1956). Good and evil. *Analysis*, *17*, 33–42.

Heim, I., & Kratzer, A. (1998). *Semantics in generative grammar*. Oxford: Blackwell.

Jespersen, B. (2005). Explicit intensionalization, anti-actualism, and how Smith's murderer might not have murdered Smith. *Dialectica*, *59*, 285–314.

Jespersen, B. (2008). Predication and extensionalization. *Journal of Philosophical Logic*, *37*, 479–499.

Jespersen, B., & Primiero, G. (forthcoming). *Two kinds of procedural semantics for privative modification.* (*Lecture Notes in Artificial Intelligence*)

Kamp, H., & Partee, B. (1995). Prototype theory and compositionality. *Cognition*, *57*, 129–191.

Partee, B. (2001). Privative adjectives: subsective plus coercion. In R. Bäuerle, U. Reyle, & T. Zimmermann (Eds.), *Presuppositions and discourse*. Amsterdam: Elsevier. (Available at http://people.umass.edu/partee/docs/ParteeInPressKampFest.pdf)

Rotstein, C., & Winter, Y. (2004). Total adjectives vs. partial adjectives: scale structure and higher-order modifiers. *Natural Language Semantics*, *12*, 259–288.

Tichý, P. (1971). An approach to intensional analysis. *Noûs*, *5*, 273–297. (Reprinted in *Collected papers in logic and philosophy*, eds. V. Svoboda, B. Jespersen and C. Cheyne, Prague: Filosofia (Czech Academy of Sciences), and Dunedin: University of Otago Press)

Gentzen's Consistency Proofs for Arithmetic

Annika Kanckos

1 Gentzen's four proofs

The limitations of the formal systems of arithmetic, revealed by Gödel's incompleteness theorems in 1931, imply that the consistency of Peano arithmetic can be established only by reducing the problem to certain fundamental principles that are not formalizable within the theory itself. Although the finite methods advocated by Hilbert's program could not succeed in producing a proof, the search continued with broader constructive methods. For a proof to be meaningful, in Hilbert's sense, the principles used should be considered more reliable than the doubtful elements of the theory concerned.

In 1933, Gentzen and Gödel independently established that the consistency of classical Peano arithmetic reduces to the consistency of intuitionistic Heyting arithmetic by negative translation. Therefore, an inconsistency in the classical calculus cannot stem from the principle of indirect proof, which by some was considered a doubtful element. The only remaining principle of arithmetic whose consequences could be doubted was induction.

The earliest proofs of the consistency of Peano arithmetic were presented by Gentzen, who worked out a total of four proofs that were published between 1936 and 1974. The first consistency proof was withdrawn from publication due to criticism by Bernays for implicit use of the fan theorem, although this assessment was later retracted (Bernays, 1970). However, a galley proof of the article was preserved and excerpts were published posthumously in English translation (Gentzen, 1969), as well as unabridged in the German original (Gentzen, 1974). The critique led to an alteration of the argument and the published second proof (Gentzen, 1936), which is appended with an ordinal assignment and relies on a constructive proof of the principle of transfinite induction up to the ordinal ϵ_0.

In both the first and the second proof, the arithmetical system is formalized in natural deduction written in sequent calculus style. The third consistency proof (Gentzen, 1938) is a revised version, conducted in pure sequent calculus, whereas the fourth proof (Gentzen, 1943), in contrast, proves consistency through a non-derivability.

The combinatorial methods of Gentzen's reduction procedure described in the third proof can be represented in primitive recursive arithmetic (PRA). PRA is a weaker theory than Peano arithmetic and it is generally included in, and often identified with, finitistic logic, because unbounded quantification over the domain of natural numbers is not allowed. Due to this feature, the primitive recursive operation on derivations, described in Gentzen's proof, corresponds to a quantifier-free formula. Therefore, finitistic reasoning together with the principle of transfinite induction restricted to quantifier-free formulas gives the consistency result. It should be noted that the theory, in which the proof is formalizable, is incomparable to Peano arithmetic. The theory is not stronger than Peano arithmeic, becauce complete induction cannot be proved. But on the other hand, neither is the theory weaker, becauce it proves the consistency of Peano arithmetic.

2 Gentzen's work related to Hilbert's programme

During the early 1900's Hilbert developed his views on the foundations of mathematics and presented his views in a succession of papers. He proposed a method for solving the foundational crisis that had emerged after the paradoxes of set theory. In 1921 the aims of Hilbert's programme consisted of formalizing all mathematical theories, and providing 'finitary' consistency proofs for them. Furthermore, the programme included that the questions of mutual independence and completeness of the axioms of the theory were to be answered and possibly a decision method found for the theory.

By Gödel's incompleteness theorem from 1931 it was shown that no formalization of elementary arithmetic can be complete and that it is impossible to find a finite consistency proof in the sense that Hilbert's programme required. Therefore, the methods that are proper to the theory, the consistency of which we are proving, do not suffice when proving consistency of the theory. To produce a consistency proof, the consistency of the methods used need to be presupposed. That is, no absolute consistency proof exists and all proofs merely reduce the question of consistency to that of the other theory used.

The incompleteness theorem means that in any formal theory, there are always true number-theoretical sentences that are not provable within the theory. Another description of the result is that "sentences can always be found, the proofs of which again always require new modes of inference". (Gentzen, 2007, p. 357) This reveals a weakness in the axiomatic method, implying that the consistency proofs must be extended whenever the proof means are extended. In 1937 Gentzen however regards the extensions not relevant in practice, because at that time no Gödel sentence of practical significance had been revealed, except for the sentence expressing consistency. In 1943 he would himself accomplish such an extension.

3 Gentzen on how consistency proofs are possible

Hilbert's programme had not been abandoned by Gentzen. After the finitistic methods had proven to be insufficient for proving the consistency of arithmetic, Gentzen continued the search for a proof. He concludes in (Gentzen, 1936) that a consistency proof is still possible and meaningful if the methods used are more reliable, even if not proper to elementary number theory.

In order to prove the consistency of arithmetic Gentzen followed the general aims of Hilbert's program, which were to prove consistency "by means of inference that are completely unimpeachable ('finitist' forms of inference)."[1] Therefore, Gentzen's work explores the consequences of the 'finitist' view of formalist mathematics as stated by Hilbert.

In a letter from 1932 Gentzen states that through the formalisation of logical deduction, the task of producing a consistency proof becomes a purely mathematical problem.[2] This was at a time when he worked on extending the consistency proof for arithmetic to include the rule of induction. In a lecture from 1937 (Gentzen, 2007, p. 355) Gentzen characterized Hilbert's pogramme as a way to reduce the metamathematical presuppositions and followed Hilbert in this respect.

---

[1] *The Collected Papers of Gerhard Gentzen*, p. 135.
[2] Eckart Menzler-Trott, *Logic's lost genius: The life of Gerhard Gentzen*, p. 29.

In his paper from 1936 Gentzen clearly states that the purpose of the proof is to reduce the question of consistency of arithmetic to certain general and fundamental principles. It is possible to reduce some parts of arithmetic to other parts, e.g. arithmetic of complex numbers to that of real numbers. But Gentzen concludes that there remains the task of proving the consistency of elementary number theory. His main concern is with the proving of the consistency of the logical reasoning used when proving statements about the natural numbers. This means that the consistency of the system of axioms or the basic relations between numbers is not what he is aiming to prove, because it is the reasoning employed that may produce antinomies. Gentzen discusses to what extent it is possible to carry out a consistency proof and claims that it is both necessary and possible to produce a proof, due to the paradoxes that had emerged in other areas of mathematics.

Gentzen's aim was to prove the consistency of classical mathematics, in the first place arithmetic and then analysis, by extending the methods to constructive or intuitionistically acceptable methods.[3] The constructive method used as a foundation for the consistency proofs is similar to, although somewhat broader than, Hilbert's 'finitist' standpoint. In Gentzen's opinion it provides a secure foundation because it employs the concept of possible infinity, not an actual infinity. The actual infinity is identified as a doubtful element in the methods of proof. The constructive concept of infinity, on the other hand, is not included in the framework of elementary number theory and is conjectured to be extensible beyond any formal theory.

These methods include a constructive interpretation of quantification over the infinite domain of natural numbers. If the numbers are substituted one by one in a formula that has been universally quantified, then the result is a true formula. An existentially quantified formula, on the other hand, means that a witness to the formula has been found. Even so, Gentzen thinks that some methods encountered in his proof 'give cause for concern' from the finitist standpoint, in particular the principle of transfinite induction. [4] Whether the proof can be regarded as finitist depends on if the principle of transfinite induction can be accepted as a finitist method.

---

[3] G. Gentzen, *The current situation in research in the foundations of Mathematics* and Jan von Plato, *From Hilbert's programme to Gentzen's programme*, p. 383.

[4] *The Collected Papers of Gerhard Gentzen*, p. 136.

It can also be noted that Tarski regarded Gentzen's proof as an interesting metamathematical result, but he did not think that the proof made the consistency of arithmetic more evident than by epsilon.[5]

In 1933 the Gödel–Gentzen negative translation showed that classical arithmetic can be reduced to intuitionistic arithmetic, implying that the constructive methods go beyond finitistic reasoning. Therefore, Hilbert's programme could be continued if it was modified to use the broader constructive methods. In the light of this, the negative translation was considered the first consistency proof for arithmetic.[6] Thus, Gentzen's consistency proofs, as well as Gödel's Dialectica interpretation can also be seen as part of the extended Hilbert programme.

4 Crisis and paradoxes

The importance of consistency proofs was debated due to the foundational crisis at the time when Gentzen published his proofs. Gentzen points out that despite the efforts to find a solution for the paradoxes of set theory, that is, to pinpoint the fallacy of the reasoning that leads to antinomies, a clear solution should not be expected. The flaw in the reasoning cannot definitely be pointed at. Gentzen however follows the proponents of intuitionism by claiming that the antinomies of set theory have their origin in the liberal use of the concept of infinity. He claims that "we can only say definitely that the materialisation of the antinomies is connected with the *concept of infinity*, because a purely finite mathematics, as far as anyone can judge, no contradictions can arise, provided the mathematics is correctly constructed." (Gentzen, 2007, p. 353)

One simple solution is to draw a clear line between permissible and impermissible modes of inference, thereby blocking the undesired inferences that lead to antinomies. This method has been employed, for example, in axiomatic set theory. However, according to Gentzen (Gentzen, 2007, p. 353) this solution gives rather arbitrary restrictions if the source of the antinomies has not been properly identified. Furthermore, new antinomies may in the future prove to be derivable with the allowed inferences.

In Gentzen's opinion Russell's paradox reveals a fault in the logical inferences involved. He opposes impredicative definitions and regards

---

[5] Eckart Menzler-Trott, *Logic's lost genius: The life of Gerhard Gentzen*, p. 81.
[6] Jan von Plato, *From Hilbert's programme to Gentzen's programme*, p. 392

only constructive definitions valid. New sets should be defined on the basis of already formed sets, because it is illicit to define an object by means of a totality and then to regard it as belonging to that totality, so that it contributes to its own definition. The definition of a set of all sets circular, as this set is defined and then concluded to belong to itself.[7] A problem that emerges from invalidating impredicative reasoning is that this form of reasoning is also used in analysis, in the proofs of basic theorems, such as the intermediate value theorem. The definition of the intermediate point is problematic, because the point is included in the intervals defined in the proof of the theorem. This means that the point is defined by referring to the totality of reals and is then concluded to belong to this totality.

A radical standpoint is taken by the intuitionists who do not consider the arguments used in classical analysis valid, because the law of trichotomy is not true on the intuitionistic continuum. Thus, they reject the means of proof that allows a division of the reals into two intervals. The intuitionist standpoint is not only taken to avoid possible antinomies, but because the classical statements are considered meaningless.

5 Bernays' critique and the fan theorem

Neither Bernays nor Gödel were satisfied with Gentzen's first consitency proof, which is shown in correspondence from Gentzen to Bernays in the fall of 1935. [8] The critique expressed that the proof made implicit use of the fan theorem. However, this claim was later retracted and it has been noted by Kreisel in 1987 that this principle is not sufficient for proving the consistency of Peano arithmetic. The principle that was implicitly used to prove termination is bar induction. Gentzen, who had already thought of the objections, reworked his proof and instead relied on the principle of transfinite induction.

König's lemma, which states that a finitely branching tree with an infinity of nodes has an infinite branch, is not constructively valid. The contrapositive of König's lemma, called the fan theorem, is however constructive. It states that if all branches of a tree are finite, then the whole tree is also finite. Bar induction is the contrapositive of a similar classical principle for infinitely branching trees.

---

[7] *The Collected Papers of Gerhard Gentzen*, p. 134.
[8] Jan von Plato, *From Hilbert's programme to Gentzen's Programme*, p. 392.

6 The principle of transfinite induction

Gentzen proved the consistency of arithmetic using the principle of transfinite induction, for which he gave a constructive proof. Let $P(\alpha)$ be a property defined for all ordinals α and let α be an arbitrary ordinal. Then if from the assumption that for all $\beta < \alpha$, $P(\beta)$ holds, follows that $P(\alpha)$ holds, then by the principle the property holds for all ordinals. His use of the principle was restricted to primitive recursive predicates. The primitive recursive predicates, $P(n)$, can be verified for an arbitrary number, n, by a bounded computation.

In his proof from 1943 he represented transfinite induction up to ϵ_0 as an arithmetical formula and showed that it is not provable in Peano arithmetic, but that any weaker induction principle is provable. In the proof the natural numbers are extended by what are called constructive ordinals. The induction principle is also extended into a transfinite induction principle.

Schütte and Schwichtenberg (Schutte & Schwichtenberg, 1990) note that "the transfinite induction certainly transcended the finite standpoint, as by Gödel is necessary, but it proceeds in a completely constructive way, so that the proof of Gentzen is seen as a testimonial for pure number theory in the sense of the extended Hilbert Programme..."

In general, a set-theoretical proof of the principle of transfinite induction is not acceptable if the methods are to be considered reliable from a constructive point of view. Instead Gentzen proves that each ordinal up to ϵ_0 is accessible. Accessibility means that all descending chains of ordinals are finite, or that the ordinals are well-ordered. This principle is used in order to prove termination in a finite number of steps of the reduction procedure described in Gentzen's proof.

7 From 1936 to 1938

The calculus used in Gentzen's first two proofs is natural deduction in sequent calculus style. The calculus has rules from natural deduction operating on sequents. Instead of left rules, operating on the antecedent, there are elimination rules operating on the succedent.

In the latter proof from 1936 the number of rules is decreased. Gentzen used initial sequents to replace logical rules and he replaced the left implication rule by allowing sequents of the form $A \supset B, A \rightarrow$

B. Gentzen regarded structural rules as purely formal modifications of the sequents, except for the rule of weakening. These rules were added to the calculus in order to obtain special features for the formalism.

The proof from 1936 can be explained as a 'reduction procedure' for sequents. Firstly, all free variables are replaced by numerals and then choices are made as the sequent is reduced to less complex sequents. The choices made can be regarded as aiming for the worst possible scenario, in which the sequent is falsified. The reduction ends in a reduced form, which consists of a true sequent. A sequent is true if the antecedent contains a false atomic formula or if the succedent is a true atomic formula. Gentzen shows that initial sequents are reducible and that the rules preserve reducibility. Consistency follows from the fact that the sequent $\rightarrow 0 = 1$ is not in reduced form nor reducible.

The proof also gives an ordinal assignment to prove that the process terminates. This can be compared to the proof from 1938, which also uses an ordinal assignment, but has a standard notation for the ordinal numbers.

As a standard version of the classical consistency proof we consider Takeuti's version (Takeuti, 1987), which is based on Gentzen's proof (Gentzen, 1938). This third proof is the best known of Gentzen's papers on this subject. Gentzen's consistency proof from 1938 can be explained as consisting of a well-ordering of all derivations and a reduction procedure for derivations of the empty sequent. Derivations are ordered by complexity and a reduction procedure is given for all derivations of the empty sequent, such that the reduction decreases the complexity of the derivation. Therefore, if there exists a derivation of the empty sequent, then by a finite number of steps a simple derivation, which does not contain any induction rule, is reached. Consistency then follows by proving that the empty sequent is not derivable without the induction rule.

In the article from 1938 a standard multi-succedent sequent calculus is used. In contrast to the earlier proofs the reduction process resembles cut elimination. In the first step of the reduction procedure free variables in a derivation of the empty sequent are replaced by numerals. Then the 'end-piece' of the derivation is considered. The end-piece consists of structural rules and induction at the end of the derivation. Induction rules and initial sequents are reduced if they occur in the end-piece. Lastly, cuts on compound formulas are reduced.

The cuts are not directly reduced to cuts on less complex formulas, but additional cuts are on the less complex formulas are introduced. This introduction of additional cuts is called the height-line argument. The ordinal assignment defines a notion of height of a cut and the additional cuts push up the places in the derivation where the heights of the cuts drop. These drops affect the ordinal assigned and the result is a reduction of the ordinal of the derivation.

8 Natural deduction and sequent calculus

In Gentzen's opinion the object of logic is to study the general structures of proofs. This opinion is a break with the logicist tradition of Frege, Peano, Russell and Hilbert who considered the object of logic to study logical truth.[9]

Gentzen developed the systems of natural deduction and sequent calculus to analyze the structure of proofs. The former was successful for the intuitionistic case and the latter was needed to deal with the classical case. Natural deduction with its hypothetical reasoning was developed to echo better than axiomatic calculi, the actual reasoning in mathematical proofs. It can be noted that the system of natural deduction was independently developed by Jaskowski in 1934. His system is presented in linear form and his work does not contain any analysis of the structure of the derivations. [10]

Sequent calculus formalizes the derivability of a formula from other formulas, $\Gamma \to C$, represented by an arrow between a list of assumptions, Γ, as an antecedent and a conclusion, C, as a succedent of the sequent. As a generalization of the notion of sequent a classical multi-succedent calculus is obtained. In the classical calculus the sequents, $\Gamma \to \Delta$, can be interpreted as a number of cases, Δ, under the open assumptions, Γ. It should be noted that it is not necessarily decidable which of the cases hold.

In Gentzen's formalized systems intuitionistic logic gains a strong position because it becomes a special case of the classical calculus. This property may not be as striking in other calculi, in which the rules are chosen differently. The calculi show his intent to use intuitionistic logic as a base for his argumentation.

---

[9] *ibid.*, p. 384.
[10] Jaskowski, *On the rules of supposition in formal logic*, (translation of polish original) in ed. S. McCall, Polish Logic 1920–1939, pp. 232–258, Oxford 1967.

9 Gentzen's consistency proof performed in natural deduction

Since the publication of Gentzen's proof, the conducting of the consistency proof in standard natural deduction has been an open problem. This problem has recently been solved for an intuitionistic calculus by the present author. The result is based on a normalization proof by Howard (Howard, 1970), recommended to the author by Per Martin-Löf. The new consistency proof is performed in the manner of Gentzen, by giving a reduction procedure for derivations of falsity. In contrast to Gentzen's proof, the procedure is appended with a vector assignment. The reduction reduces the first component of the vector and this component can be interpreted as an ordinal less than ϵ_0, thus ordering the derivations by complexity and proving termination of the process.

One of the main differences of how the derivations are treated is in the reduction of an instance of the rule of induction in the derivation. The natural deduction system produces a non-normality in the derivation as it introduces an implication, which is directly followed by an elimination of the same implication. If Gentzen's 1938 proof were translated into natural deduction, the reduced implication would become a composition of the premises of the induction rule.

10 Following in Gentzen's path

Through Gentzen's work ordinal analysis became known. This is the method of measuring the proof-theoretic strength of a formal system of mathematics, by the least ordinal with the property that no recursive well-ordering of ordinal type α may be proven well-ordered in the system in question. The proof-theoretical ordinal of first-order arithmetic is ϵ_0, which was proven by Gentzen in his proof from 1943.

After the Second World War some subsystems of classical analysis were proven to be consistent using the methods developed by Gentzen. By restricting the application of the comprehension axiom in second order predicate calculus, subsystems of classical analysis are obtained. For some of these systems it is possible to produce constructive consistency proofs.[11]

---

[11] *Ein Jahrhundert Mathematik 1890–1990 Festschrift zum Jubiläum der DMV*, p. 725.

Annika Kanckos
Department of Philosophy, University of Helsinki
P.O. Box 24 (Unioninkatu 40 A), FI — 00014 University of Helsinki
Finland
annika.kanckos@helsinki.fi
http://www.helsinki.fi/filosofia/filo/henk/Kanckos.html

References

Bernays, P. (1970). On the original gentzen consistency proof for number theory. *Intuitionism and proof theory*.

Gentzen, G. (1936). Die widerspruchsfreiheit der reinen zahlentheorie. *Mathematische Annalen*, *112*, 493–565.

Gentzen, G. (1938). Neue fassung des widerspruchsfreiheitsbeweises fur die reine zahlentheorie. *Logik und zur Grundlegung der exakten Wissenschaften*, *4*, 19–44.

Gentzen, G. (1943). Beweisbarkeit und unbeweisbarkeit von anfangsfallen der transfiniten induktion in der reinen zahlentheorie. *Mathematische Annalen*, *119*, 140–161.

Gentzen, G. (1969). *The collected papers of gerhard gentzen* (M. Szabo, Ed.). Amsterdam: North-Holland.

Gentzen, G. (1974). Der erste widerspruchsfreiheitsbeweis fur die klassische zahlentheorie. *Archiv fur mathematische Logik*, *16*, 97–118.

Gentzen, G. (2007). The current situation in research in the foundations of mathematics.

Howard, W. A. (1970). Assignment of ordinals to terms for primitive recursive functionals of finite type. *Intuitionism and proof theory*.

Schutte, K., & Schwichtenberg, H. (1990). Mathematische logik. *Ein Jahrhundert Mathematik 1890-1990. Festschrift zum Jubilum der DMV*.

Takeuti, G. (1987). *Proof theory*. Amsterdam: North-Holland.

What is natural about natural deduction
John T. Kearns

1 Making and discharging hypotheses

I think that certain deductive systems were first called *natural* to contrast them with axiomatic deductive systems, either with or without a rule of substitution. The axiomatic systems were designed to establish results that were single formulas, or single sentences, and these "logical laws" were inferred directly from other logical laws. The systems of *Principia Mathematica* were like this, as are the systems presented in Hilbert and Ackermann's *Principles of Mathematical Logic* and Church's *Introduction to Mathematical Logic*. When compared to deductive reasoning carried out using expressions of natural language, these logical deductions have a somewhat "contrived" character.

Systems of natural deduction, in contrast, have arguments, or deductions, or proofs which begin with *hypotheses*, and infer consequences of these hypotheses. A system of natural deduction might establish results linking sets of premises to conclusions; these would not be single-formula or single-sentence results. However, systems of natural deduction typically employ rules (inference principles) which *discharge*, or *cancel* hypotheses. For example, this argument from three hypotheses to the conclusion 'C':

$$\cfrac{B \qquad \cfrac{A \qquad [A \supset [B \supset C]]}{[B \supset C]} \text{Modus Ponens}}{C} \text{Modus Ponens}$$

might be continued, using a rule $\supset$ *Introduction*, to obtain an argument from '$[A \supset [B \supset C]]$' to this conclusion: $[B \supset [A \supset C]]$. A system of natural deduction can establish single-formula or single-sentence results. A particular system might establish only single-formula or single sentence results.

It is essential to a system of natural deduction that it contain proofs, or arguments, from hypotheses which need not themselves be logical laws. It is also essential, if a natural deduction system is to accommodate arguments that are commonly made when using expressions of natural language, that the system contain rules which discharge hypotheses. We might show that a statement A is true, for example, by deriving a contradiction from the hypothesis $\sim A$. We discharge the negative hypothesis when we accept A. When making an inference which discharges a hypothesis, we are actually using a whole (sub-)argument as a premiss. The discharged hypothesis is "confined" to the subargument premiss.

Natural-deduction systems typically employ introduction and elimination rules which involve occurrences of a single operator. There might be the rules $\supset$ *Elimination*, or *Modus Ponens*, and $\vee$ *Introduction*:

$\supset$ *Elimination* $\qquad\qquad\qquad$ $\vee$ *Introduction*

$$\frac{A \quad [A \supset B]}{B} \qquad\qquad \frac{A}{[A \vee B]} \quad \frac{B}{[A \vee B]}$$

From this perspective, *Modus Tollens*:

$$\frac{[A \supset B] \quad \sim B}{\sim A}$$

would not be a suitable rule, because it "involves" two connectives rather than just one. A one operator rule makes clear the contribution of that one operator, but with two operators, it isn't so clear what role each operator plays. However, it is not absolutely essential that a system of natural deduction employ one-operator introduction and elimination rules. A system of natural deduction might, after all, have *Modus Tollens* as one of its rules.

Arguments, or deductions, in a system of natural deduction resemble deductive arguments or proofs in fields such as science or mathematics. To establish a conditional result that if A, then B, we might begin by supposing A, and then, by a deduction employing one or more steps, reason to the conclusion B. This would normally be taken as sufficient to establish the conditional result, especially if the result to be established were stated before deducing B from A. The arguer isn't required to *finish* her deduction by saying, "Hence, if A, then B." But what she isn't required to say is nevertheless tacitly understood. The supposition of A is discharged once it has been shown that if A, then B.

Someone who begins an argument (proof) by supposing that $\sqrt{2}$ is a rational number, and then deduces a contradiction, might conclude by saying, "So $\sqrt{2}$ isn't a rational number," which cancels the initial supposition. If she had announced the result before giving the argument, she might not repeat that result, but could easily say something like "QED" after reaching the contradiction. With the proof by contradiction, she would probably say something to indicate that the conclusion has been established. However, this isn't the case with all arguments that discharge hypotheses. When establishing a conditional, or a universal claim, or even the consequence of a disjunction, once the argument from hypotheses which need to be discharged is completed, it is taken to be evident that the appropriate result has been established, and this is often, perhaps usually, not stated explicitly. If we begin by letting ABC be a triangle, and eventually conclude that the sum of the interior angles of this triangle is equal to two right angles, we can simply stop, and our audience will know that our result holds for all (Euclidean) triangles.

2 Illocutionary acts

We commonly do carry out deductive reasoning from hypotheses in natural languages, where the hypotheses must be discharged in order to establish results. We find this natural. People reasoned in this way before the development of formal logical theories. Systems of natural deduction formalize a kind of reasoning that has long been familiar in deductive sciences. Although it is widely recognized that arguments from hypotheses are a natural form of reasoning, certain features of these arguments are frequently not recognized. We commonly mark a hypothesis in an argument by using either the word 'suppose' or the word 'let' (as in "Let ABC be a triangle"). I shall call these hypotheses *suppositions*; they are *initial suppositions* of deductive arguments.

Suppositions are most appropriately compared with assertions, or judgments, and denials. Assertions and denials are *illocutionary acts*. A statement can be accepted (as representing, things as they are) or rejected. Considering a statement and coming to accept it is an act. Once a person has performed this act, she continues to accept the statement until she either changes her mind, or forgets that she has accepted the statement. Continuing to accept a statement is a state rather than an act. An assertion is either an act of coming to accept a statement, or an act reaffirming one's continued acceptance of the statement. Assertions are typically directed by a language user to an addressee. But I am using this word to cover acts of producing a statement, and accepting or reaffirming it, whether or not there is an audience. An assertion can involve a spoken statement, or a written statement, or even a statement which is merely thought. A denial is an act of rejecting a statement as one not suitable for being asserted. A *statement* is a *speech act*, or *language act*, which is true or false, and which represents things as being this or that.

A statement can be asserted or denied, it can also be (positively) supposed to be the case — this is to accept it temporarily, or provisionally. However, it isn't only statements that can be supposed to be the case — we can also do that with schematic sentential expressions like 'ABC is a triangle.' A statement or schematic language act can also be temporarily or provisionally rejected. On this understanding, if we make a supposition, and infer a consequence of the supposition, our conclusion also has the status of a supposition, and will be called (by me) a supposition. Like assertions and denials, suppositions are illocutionary acts.

We can distinguish *initial* suppositions from *dependent* suppositions, which are derived from, and depend on, other suppositions. An assertion or denial can be an initial assertion or denial in a particular argument, but if, say, an assertion is derived from other illocutionary acts, it doesn't depend on those acts. We can simply accept the asserted statement, and disregard the premises from which it was derived.

A genuine deductive argument is a speech act, and its basic components are illocutionary acts. The argument might begin with assertions, denials, and suppositions, and proceed to a conclusion which is also an act of one these kinds. To understand, investigate, and evaluate speech-act arguments, we need to pay attention both to truth conditions of statements and to the illocutionary force with which statements are made. Standard systems of logic, or logical theories, fail to do this. Standard deductive systems do not "provide" for making, or representing, speech–act arguments. Instead, they license what I shall call *deductive derivations*: these are concerned solely to investigate truth conditions, and to trace truth-conditional "connections" among statements. It is easiest to see this in an axiomatic deductive system where both axioms and theorems are logical laws or (if they are sentences) logical truths.

What is natural about natural deduction

Even standard systems of natural deduction, though their derivations mimic natural arguments to a certain extent, are exclusively focused on truth-conditional connections. These systems don't have a notational device for indicating illocutionary force, and so lack the resources to recognize or give an account of certain features that are essential to the correctness of a deductive argument. Imagine that we have a formal language with expressions for representing true or false statements, and, perhaps, with schematic sentential expressions. Then let us introduce the expressions for indicating illocutionary force. These are prefixed to the sentential expressions:

$\vdash$ the sign of assertion $\quad\dashv$ the sign of denial
$\llcorner$ the sign of supposing true $\quad\neg$ the sign of supposing false

A sentence in a language of propositional logic that is composed exclusively from atomic sentences and conventional logical operators is a *plain sentence* of the logical language. The result of prefixing a plain sentence with an illocutionary operator is a *completed sentence*. Plain sentences represent statements, while completed sentences represent illocutionary acts.

A derivation like the following:

$$\cfrac{\cfrac{\cfrac{[A\&[B\&C]]}{[B\&C]}\&E}{B}\&E \quad \cfrac{[D\&E]}{D}\&E}{[B\&D]}\&I$$

can enable a person to determine that if the hypotheses '$[A\&[B\&C]]$' and '$[D\&E]$' are true, then '$[B\&D]$' must also be true. In order to modify this derivation so that the resulting construction fully represents a genuine speech-act argument, the expressions on each line must be prefixed with an expression which indicates illocutionary force.

One way of doing this yields the following representation:

$$\cfrac{\cfrac{\cfrac{\vdash[A\&[B\&C]]}{\vdash[B\&C]}\&E}{\vdash B}\&E \quad \cfrac{\vdash[D\&E]}{\vdash D}\&E}{\vdash[B\&D]}\&I$$

This represents an argument in which every step is an assertion. An argument having this form is, evidently, deductively correct. (It isn't appropriate to characterize such an argument as being valid or invalid, as those expressions are customarily understood.) It is important to realize that this representation, and the argument that is represented, is essentially *first-person*. The *illocutionary operators* indicate the illocutionary acts being performed by *the person who is making the argument*.

If the steps in the original representation are prefixed with different illocutionary operators, as in the following:

$$\cfrac{\cfrac{\cfrac{\vdash [A\&[B\&C]]}{\vdash [B\&C]}\&E}{\vdash B}\&E \quad \cfrac{\cfrac{\llcorner [D\&E]}{\llcorner D}\&E}{\vdash [B\&D]}}{\&I}$$

the argument that is represented is not deductively correct. The last "move" is the one that is mistaken. A premiss that is asserted, when combined with a positive supposition, will not support an asserted conclusion. For an argument like this (an argument whose initial acts are all positive) to be deductively correct, any way of satisfying the truth conditions of the initial statements must also satisfy the truth conditions of the statement in the conclusion, and, in addition, the illocutionary force of the conclusion of an argument must not exceed the force of the premisses.

In a system of illocutionary logic, at least in the kind of system which I'm promoting here, the deductions are perspicuous representations of the kind of speech act arguments that are actually performed outside of logical studies. They make it convenient to distinguish considerations of truth conditions from those involving illocutionary force, and should prove useful for investigating features like cogency and rigor.

3 Illocutionary semantics

It is to some extent arbitrary how we "demarcate" the field to be called semantics. Semantics might be construed as the study of truth conditions of statements. It can be construed more broadly than this. If semantics is confined to truth conditions and features that can be defined in terms of truth conditions, then what I have called deductive derivations pretty much exhaust the deductive techniques for exploring semantic concepts.

I favor a broader understanding of what is included in semantics. On my conception, both truth conditions and illocutionary force are semantic features of speech acts. Statements are those sentential acts which have truth conditions, and they are considered in abstraction from illocutionary force. Illocutionary acts are constituted by sentential speech acts performed with a certain force. The "contents" of assertions, denials, and suppositions, among others, can be regarded as statements. Not every sentential speech act is a statement or contains a statement. In a system of illocutionary logic which deals with arguments like those illustrated earlier, we need to recognize a semantic feature or features of illocutionary acts in terms of which we can characterize deductively correct arguments.

I think the appropriate feature is the one I call *rational commitment*. This feature, when recognized, can motivate a person to perform an intentional act. Making a decision to carry out a given action rationally commits a person to carry out that action. Performing some intentional acts can commit a person to perform, or not perform, other intentional acts. We can also be committed to remain in a certain state, like that of accepting a given statement. But rational commitment need not involve a moral requirement. I can decide to do something like buy gas on the way to work, and then fail to do it, either because I forget what I intended to do, or for some other reason, without being culpable in any way.

Some commitments are conditional, like the commitment to close the windows upstairs if it rains while I am at home, and others, like my commitment to buy gas on the way to work, are unconditional. Coming to accept, or continuing to accept, some statements, while rejecting other statements, will commit a person to accepting further statements, and to rejecting further statements. Positively or negatively supposing statements will commit a person to supposing others (either positively or negatively). If the person who accepts certain statements, and rejects others, is committed to, say, accept statement A, this commitment is conditional. She is committed to accept A if she has some interest in the matter, and gives it some thought. Although I accept many statements and reject many others, I am not interested in exploring the consequences of most of these beliefs and disbeliefs. However, it is irrational to accept a disjunction "A or B," reject B, but refuse to accept A.

Implication relations among statements are ontological, or *ontic*, features, not epistemic ones. A group of statements either implies some further statement or it doesn't. Some implications are more difficult to recognize, or grasp, than others, but the complicated cases of implication are not composed of, or constituted by, simpler cases. However, with rational commitment, which is an epistemic feature, there is an important difference between *immediate* commitment and *mediate* (or remote) commitment. Immediate commitment is evident to the person for whom it is immediate. If doing A immediately commits a person to doing B, and doing B immediately commits her to doing C, then doing A may only mediately commit her to doing C. It is immediate commitment which, when recognized, motivates a person to act.

A given person, at a time, is characterized by her commitments to perform or not perform certain acts, including her commitments to accept or reject certain statements. Commitment is a natural feature of the human landscape. In a genuine deduction, a *natural* deduction, a person traces the immediate commitments of her illocutionary acts and of the states that her acts reflect. A truly natural deduction involves moves from illocutionary acts to illocutionary acts, and the correctness of such a deduction depends on both truth conditions and illocutionary force. But the commitment relation linking illocutionary acts and their associated states to further acts and states is constituted by a combination of illocutionary force and truth conditions.

Since commitment involves doing or not doing certain things, or continuing in a state that is intentionally entered into, the most appropriate way to present, and to explain, deductive commitments is by presenting deductive principles and developing a deductive system. The very idea of commitment in a system of illocutionary logic is best captured by characterizing deductions. But deduction and deductive systems are not semantic, or are not conceived as semantic. A logical semantic account employs functions which assign things to expressions, so that the values of complex expressions are partly or wholly determined by the values of their components. In a standard system of logic, or the ontic component of a system of illocutionary logic, the semantic account seems to capture the truth-conditional meanings of logical expressions. The deductive system, which is a system of what I have called deductive derivations, codifies certain semantically distinguished expressions of the formal language.

In a system of illocutionary logic, at the epistemic level, it is the deductive system that best captures the idea of commitment, but a semantic account which assigns values to completed sentences can be provided. I consider this account to be semantic because of the techniques employed in presenting and exploring it, not because this account captures meaning in some intuitive sense. The semantic account is initially a conjecture, which is established once the deductive system is shown to be sound and complete with respect to the semantic account.

To develop a semantic account for a system of illocutionary logic, we employ two kinds of function. The first kind of function, interpreting functions, determine the truth and falsity of plain sentences of the logical language in pretty much the standard way. In a system of propositional logic, for example, an interpreting function assigns truth and falsity to the atomic plain sentences of the language, and this determines a valuation of all the plain sentences. In a system of first-order logic, the interpreting function of the language for a domain assigns values to individual constants and predicates, and this determines a valuation of all the plain sentences.

The second kind of function assigns values to (some of) those completed sentences which are either assertions and denials. Since certain suppositions depend on others, we don't employ functions which assign values to suppositions. For assertions and denials, we consider an idealized language user, a woman whom I call the *designated subject*, at some particular time. At this time, there are statements which she has accepted and continues to accept, and statements which she has rejected and continues to reject. These *explicit beliefs* and *disbeliefs*, at that time, commit her to accept further statements, and to reject further statements. I use the plus sign '+' for the value of assertions and denials that she has already performed or that she is committed to perform at that time.

A *commitment valuation* is a function which assigns the value + to some completed sentences $\vdash A$ and $\dashv B$ of the logical language. We say that a commitment valuation V is *based on* an interpreting function f iff (i) for every completed sentence $\vdash A$ which is assigned + by V, the plain sentence A has value T for the valuation determined by f, and (ii) for every completed sentence $\dashv B$ which is assigned + by V, the plain sentence B has value F for the valuation determined by f.

A *coherent* commitment valuation is one that is based on an interpreting function. Someone whose beliefs and disbeliefs are characterized by a coherent commitment valuation is a person whose beliefs and disbeliefs dont conflict — the beliefs might all be true, and the disbeliefs false. If commitment valuation V is based on interpreting function f, then $\langle f, V \rangle$ is a *coherent pair*.

If a coherent commitment valuation V_0 characterizes the designated subjects explicit beliefs and disbeliefs at a given time, this valuation determines a further commitment valuation, the *completion* of V_0, which assigns $+$ to those assertions and denials to which the designated subject is committed by her explicit beliefs and disbeliefs. To provide a formal characterization of the completion of a coherent commitment valuation, it is helpful to consider a restricted class of interpreting functions, the *admissible* interpreting functions. Then we characterize the completion as follows:

Let V_0 be a coherent commitment valuation. The *completion* of V_0 is the commitment valuation V such that (i) $V(\vdash A) = +$ iff A has value T for every admissible interpreting function on which V_0 is based, (ii) $V(\dashv B) = +$ iff B has value F for every admissible interpreting function on which V_0 is based.

If the person who performs the illocutionary acts represented by completed sentences $A_1, \ldots, A_n$ must be committed to perform the act represented by B, we say that $A_1, \ldots, A_n$ *logically require* B. To formally characterize this relation, we use these definitions:

Let A be a completed sentence of the logical language, V_0 be a coherent commitment valuation, and V be the completion of V_0. Then V_0 *satisfies* A iff $V(A) = +$.

Let A, B be plain sentences of the logical language, and f be an interpreting function. Then (i) f *satisfies* $\vdash A$ iff $f(A) = T$, and (ii) f *satisfies* $\neg B$ iff $f(B) = F$.

Let A be a completed sentence of the logical language, and $\langle f, V \rangle$ be a coherent pair. Then $\langle f, V \rangle$ *satisfies* A iff (i) A is an assertion or denial, and V satisfies A, or (ii) A is a supposition, and f satisfies A.

Now we can say that completed sentences $A_1, \ldots, A_n$ *logically require* (completed sentence) B iff (i) B is an assertion or denial, and every coherent commitment valuation which satisfies the assertions and denials among $A_1, \ldots, A_n$ also satisfies B, or (ii) B is a supposition, and every coherent pair which satisfies all of $A_1,\ldots,A_n$ also satisfies B.

Just as we have an epistemic-level concept *logical requiring* which is a counterpart to the ontic-level concept *implication*, so we have a concept *incoherence* which is a counterpart to (semantic) *inconsistency*:

Given a plain sentence A, these suppositions are (*logically*) *incoherent*: $\llcorner A$, $\neg A$, and these completed sentences are (*logically*) *incoherent*: $\vdash A$, $\dashv A$. Completed sentences which logically require incoherent sentences are themselves *incoherent*.

4 Distinctive features of illocutionary concepts

If statements (plain sentences) $A_1,\ldots,A_n$ imply B, then the assertions $\vdash A_1,\ldots,\vdash A_n$ logically require $\vdash B$, and the suppositions $\llcorner A_1,\ldots,\llcorner A_n$ logically require $\llcorner B$. So if a single sentence A implies B, then $\vdash A$ logically requires $\vdash B$. But if a completed sentence $\vdash C$ logically requires $\vdash D$, we cannot say that C must imply D.

To see why not, consider the first person belief operator $\boldsymbol{B}$. It is evident that the premises of the inferences below logically require the conclusions:

$$\frac{\vdash A}{\vdash \boldsymbol{B}A} \qquad \frac{\vdash \boldsymbol{B}A}{\vdash A}$$

But it isn't the case that A implies $\boldsymbol{B}A$, or that $\boldsymbol{B}A$ implies A.

Consistency also fails to line up with coherence. A sentence '$[A\&{\sim}\boldsymbol{B}A]$' which is true is consistent, no matter who the (first) person in question is. But this sentence cannot coherently be accepted/asserted by anyone. For the assertion '$\vdash[A\&{\sim}\boldsymbol{B}A]$' logically requires both $\vdash \boldsymbol{B}A$ and $\vdash{\sim}\boldsymbol{B}A$.

The divergence beween implication and logical requiring, and between consistency and coherence, is responsible for puzzles like Moore's Paradox and the Surprise Execution Paradox. By focusing on illocutionary acts in addition to statements, and the logical features of both statements and illocutionary acts, we can systematically explore the differences between the truth-conditional semantic features and their epistemic counterparts. However, I judge the most important feature of illocutionary logic to be that we can develop logical systems to capture, explain, and understand various features of our linguistic and deductive practice. (Some papers that present systems of illocutionary logic are listed in the *References*.)

John T. Kearns
Department of Philosophy and Center for Cognitive Science
University at Buffalo, SUNY
Buffalo, New York 14260
kearns@buffalo.edu

References

Kearns, J. (2000). An illocutionary logical explanation of the surprise execution. *History and Philosophy of Logic*, *20*, 195–214.

Kearns, J. (2003). The logic of coherent fiction. In T. Childers & O. Majer (Eds.), *The logica yearbook 2002* (pp. 133–146). Prague: Filosofia.

Kearns, J. (2006). Conditional assertion, denial, and supposition as illocutionary acts. *Linguistics and Philosophy*, *29*, 455–485.

Kearns, J. (2007). An illocutionary logical explanation of the liar paradox. *History and Philosophy of Logic*, *28*, 31–66.

Kearns, J. (2008). Illocutionary origins of familiar logical operators. In M. Peliš (Ed.), *The logica yearbook 2007* (pp. 55–66). Prague: Filosofia.

Descriptive Indexicals and the Referential/Attributive Distinction

Katarzyna Kijania-Placek*

1 Introduction

This paper is a contribution both to the discussion of descriptive uses of indexicals, initiated by (Nunberg, 1993), (Nunberg, 2004) and (Recanati, 1993), (Recanati, 2005), as well as to the debate about the distinction between referential and attributive readings of definite descriptions (Donnellan, 1966), (Wettstein, 1981), (Kripke, 1977), (Neale, 1990), (Devitt, 2004). Descriptive uses of indexicals are uses that generate general propositions instead of singular propositions, the latter being typically expressed by means of indexicals. To explain this usage, I will provide two examples. Imagine somebody uttering (1) on the last full day of LOGICA:

(1) Today is always the biggest party night of the conference.[1]

Or consider someone gesturing towards John Paul II as he delivers a speech with a Polish accent shortly after his election.

(2) He is usually an Italian, but this time they thought it wise to elect a Pole.[2]

We are concerned here with such uses of indexicals whereby the proposition expressed is not a singular one about a person or the day of utterance, but a general one about whomever/whatever possesses a certain property. For that, however, we need a concept word that would contribute to the general proposition, while the indexical provides us only with an object.

---

*This paper draws on research done with the assistance of the Polish Ministry of Science and Higher Education Grant N N101 132736. I am indebted to Lionel Shapiro and to other participants of LOGICA09 for valuable comments.

[1] This is a modified version of Nunberg's example (Nunberg, 1993, p. 29).

[2] This example is used by Recanati (Recanati, 2005, p. 297), but is attributed to Nunberg.

Intuitively we know what we are looking for: "today" serves as a sort of abbreviation for the "last day of LOGICA" and "he" for "Pope". Thus the question with which we are concerned is the following: From where do the properties that figure in the general propositions come, and what role does the indexical play in determining these properties.

In this paper I will not discuss Nunberg's and Recanati's answers to the question, because that is a subject to which I devote a separate paper (Kijania-Placek, 200x). I will just briefly sketch my own proposal and show how it influences the referential/attributive debate.

2 The proposal

I propose treating descriptive uses of indexicals as a special kind of anaphoric uses which I call quasi-anaphoric. In the quasi-anaphoric mechanism, an indexical expression inherits its semantic properties from its antecedent, but — in contrast to ordinary anaphora — that antecedent is derived from the extra-linguistic context. Consider the first example:

(1) Today is always the biggest party night of the conference.

From a semantic point of view, today refers to a particular day. Always, on the other hand, is a general quantifier that quantifies over days and as such requires a range of days. It is as if we were saying "today everyday" — "today" presupposes singularity while the quantifier excludes it, thus, semantically, this sentence is inconsistent.

In general, quasi-anaphora is triggered at the level of linguistic meaning by the use of quantifying words such as 'traditionally', 'always', or 'usually', whose meaning in some cases clashes with the singularity of the default referential reading of indexicals. As a result we look for discourse antecedents of the pronouns and those are the properties of the individuals in question, properties which are salient in the context. The context must be very specific to supply just one such property, which explains why there are not many convincing examples of felicitous uses of descriptive indexicals.

What is important is that the property is not a referent for the indexical, but serves as a context set that limits the domain of quantification for the quantifier that triggered the mechanism. And it is the quantifier that constrains the structure of the general proposition. Always is a binary quantifier:

Descriptive Indexicals and... 135

$$\text{Always}_x(\phi(x), \psi(x)),$$

where ϕ is the context set that comes from the extra-linguistic context. Thus, in this case we have

$\text{Always}_x(\text{last day of LOGICA}(x),$
  the biggest party night of the conference$(x))$

with the usual truth conditions

$$\mathfrak{M} \models \text{Always}_x(\phi(x), \psi(x))[g, i] \text{ iff } \phi^{\mathfrak{M}xgi} \subseteq \phi^{\mathfrak{M}xgi}$$

where g is an assignment and i is a context. In both examples there is an inconsistency between the default singularity of the indexical and the generality of the quantifier, so no singular propositions are expressed. As a result, the general propositions generated quasi-anaphorically are the propositions expressed and not just the propositions implicated in the Gricean sense (Grice, 1989).

3   A different type of descriptive uses of indexicals

Yet sometimes quasi-anaphora generates only implicated propositions. This happens when there is no semantic inconsistency and a singular proposition is expressed, but comes into conflict with the pragmatic purpose of expressing it. It is then this conflict that triggers quasi-anaphora. An interesting example was given by Nunberg.[3] It comes from the movie "The Year of Living Dangerously" by Peter Weir, in which Mel Gibson plays a reporter in Indonesia, Mr. Hamilton, who is after arms shipments for local communists. Of course, he would be in trouble if they discovered his purpose. Hamilton, talking to a warehouse manager and inquiring after the shipments, gets a warning:

— MR. HAMILTON?
BE CAREFUL WHO YOU TALK TO ABOUT THIS MATTER.
I'M NOT P.K.I., BUT I MIGHT HAVE BEEN.

---

[3] As reported by Recanati (Recanati, 1993, p. 301), this example was used by Nunberg at his lecture "Indexicality and context", which was delivered in June 1990 at the *Philosophy and Cognitive Science* conference in Ceresy-la-Salle, France.

"P.K.I." is an abbreviation for "Partai Komunis Indonesia". Let's represent what he said by

(3) I might have been a communist.

This sentence is semantically consistent and expresses a particular proposition in this context. Since indexicals always take wide scope in modal contexts, the proposition is true if and only if the speaker is a communist in some counterfactual world. This proposition is impotent as a warning, however. For Hamilton's safety here, it is totally irrelevant who his current interlocutor is in a counterfactual world. The speaker is not a communist in the actual world and this is the only thing that matters. It is in this world that somebody must be a communist in order for Hamilton to be in danger. The sense of the warning which is communicated here is not the proposition expressed but some other proposition that is quasi-anaphorically implicated. In this case it is the conflict at the pragmatic level — between the proposition expressed and the purpose of expressing it — that triggers the quasi-anaphoric interpretation. We search the context for extra-linguistic discourse antecedents for "I". It is a property of the speaker that is salient in this context, be that an 'Indonesian warehouse manager' or the 'person you talk to in Indonesia'. The property serves the purpose of the context set for the binary existential quantifier which is implicit in this kind of modal sentence. As a result we obtain first the proposition:

$\Diamond$ there exists$_x$(Indonesian warehouse manager(x), communist(x)).

This proposition is true if and only if there is a possible world in which there exists an Indonesian warehouse manager who is a communist. But, in fact, the counterfactual worlds do not matter for the warning, so the proposition implicated is simply:

There exists$_x$(Indonesian warehouse manager(x), communist(x))

meaning "There are Indonesian warehouse managers, who are communists".

As we have seen, the mechanism of quasi-anaphoric interpretation can generate either a proposition expressed or just a proposition implicated. The former is triggered by semantic inconsistency between the default interpretation of an indexical and a quantifier; the latter comes about when there is inconsistency between the proposition expressed and the pragmatic purpose of expressing it.[4]

4 Referential/attributive

What has all this to do with the referential/attributive distinction? The ongoing debate was initiated by Donnellan's 1966 paper, which contained examples similar to the following:[5]

(α) Her husband is kind to her.

uttered by somebody pointing to a woman's lover and saying truly that he is kind to her. Nobody has contested the fact that (α) can be used to say something true about the lover. The debate is about the analysis of such uses of definite descriptions. Referentialists (Devitt, 2004), (Wettstein, 1981) postulate the semantic ambiguity of the definite description, while Russellians (Grice, 1989), (Neale, 1990) explain such linguistic facts in pragmatic terms.

According to referentialists, the sentence "Her husband is kind to her" can be used to expressed two propositions depending on which of the two meanings of "her husband" is deployed: a singular one about the lover (referential reading of "her husband") or a general one about whoever is the husband (attributive reading). According to the Russellians, the same general proposition about her husband is expressed in both cases, while the singular proposition about the lover is being pragmatically implicated.

But what is the nature of the referential readings of definite descriptions? What is their status for referentialists? Are they referential terms semantically on par with indexicals? This seems to follows from Stephen Neal's Hypothesis (NH). He claimed that in natural language there are only two kinds of noun phrases: quantifiers and referential terms. (Neale, 1993) Since referentially used definite descriptions are not quantifiers, they seem to share their semantics with other referring expressions.

---

[4]In my account (Kijania-Placek, 2009), (Kijania-Placek, 200x), one other type of descriptive uses of indexicals is distinguished, but the details do not bear on the present discussion.

[5]This example comes from (Kripke, 1977) and is a modified version of one of Linsky's examples (Linsky, 1963).

5 Indexicals versus referential descriptions

But there is a difference between indexicals and referentially used definite descriptions: while indexicals are susceptible to descriptive readings, definite descriptions are not. Consider the following definite description:

Alice's eighth birthday

I may use the description referentially, having a particular day in mind, and refer to the same day, in a suitable context, by 'today'. But even though I may utter:

Today we invite lots of children every year

expressing a general proposition about Alice's birthdays in consecutive years, I cannot felicitously utter

* *For Alice's eighth birthday we invite lots of children every year.*

Although I refer to the exact same day, Alice cannot have an eighth birthday each year.

The case of propositions implicated quasi-anaphorically does not serve us any better in this test. In the movie, a first person pronoun was used by the speaker, but imagine a third person in the scene. He might have warned Hamilton by using a third person pronoun descriptively:

He might have been a communist

while pointing to the warehouse manager. But the speaker in that context could just as well refer to the manager with the help of a referentially used definite description, for example, 'the manager of this warehouse' in

The manager of this warehouse might have been a communist.

Here a particular proposition is expressed but the utterance seems pointless. After all, the manager has just said that he is not a communist. Nothing is implicated here. So it is a linguistic fact that indexicals are susceptible to descriptive interpretation, while referentially used definite descriptions are not.[6]

---

[6] On this matter, see (Nunberg, 1993), (Nunberg, 2004).

Nunberg considered explaining this difference in terms of Gricean maxims — the general proposition would always be only implicated by descriptive uses of indexicals. This assumes, however, a preceding generation of the proposition expressed. Even ignoring the difficulties that result from the inconsistency of those propositions, we must agree with Nunberg that, in this implicational interpretation, the same singular proposition would be expressed in the case of indexicals and referentially used definite descriptions, so the same propositions should be implicated.

Nunberg (in (Nunberg, 1993) and (Nunberg, 2004)) considered deploying the maxim of manner to explain the difference: even though the same proposition is being expressed, it is expressed differently and that could explain why in the case of indexicals a general proposition is implicated, while in the case of definite descriptions no proposition is implicated. But this, as Nunberg has pointed out, just cannot be explained by the maxim of manner, as this maxim explains implicatures that arise by uses of untypical expressions in contexts where typical expressions are available. Yet the use of indexicals is typical. So, if anything, the maxim of manner could explain the lack of implicature in the indexical case and its presence in the case of descriptions. That's exactly the opposite of what we need.

By proposing a quasi-anaphoric interpretation of the descriptive uses of indexicals, I am in a position to explain why indexicals are susceptible to descriptive interpretation, while definite descriptions are not. The quasi-anaphoric interpretation is blocked for referentially used definite descriptions because, even though the descriptive content of the description is truth-conditionally irrelevant, it is not transparent, as it is in the case of indexicals. The descriptive content supplies a property to the context of utterance — the property of being Alice's eighth birthday, or of being the manager of this warehouse — that blocks the salience of any other, especially more general property and thus blocks the mechanism of quasi-anaphora.

6 Nunberg's argument

Now Nunberg (in (Nunberg, 2004)) seems to be applying the case of descriptive usage of indexicals as an argument against the ambiguity treatment of definite descriptions. His argument may be reconstructed as follows, taking on the form of reductio. For his line of argument to work, however, he must assume, or rather present the referentialists as assuming something like Neal's Hypothesis (NH).[7]

ASS 1. Referentialist claim: referentially used definite descriptions are semantically different from quantifiers.

NH 2. Since they are not quantifiers, semantically they are on par with indexicals.

3. They are used to express the same singular propositions that would be expressed by the use of an appropriate indexical.

4. Since in both cases the same proposition is expressed, there should be no difference in the pragmatic properties of these expressions.

5. There is a difference in pragmatics: referentially used definite descriptions are not susceptible to descriptive readings analogous to the descriptive uses of indexicals.

6. Contradiction (4/5) ⇒ It is not the case that (1).

Thus it seems that Nunberg has shown that definite descriptions are not semantically ambiguous after all. I doubt the soundness of this argument and will therefore contest (2) together with NH.

7 A third option

My analysis of the descriptive uses of indexicals allows for a third option: definite descriptions may be ambiguous but still there is room for a semantic difference between referentially used descriptions and indexicals. On this construal, indexicals can be used to express either singular or general propositions depending on whether they are used demonstratively or descriptively. But the semantics of referentially used descriptions may be hybridous: referentially used descriptions may express quasi-singular propositions.[8] Thus their descriptive content would be truth-conditionally irrelevant – that is what they would have in common with indexicals. However, what would

---

[7] Explicitly, Nunberg assumes the hidden indexical theory of referential descriptions.

[8] In my usage of 'quasi-singular proposition' I do not follow Schiffer, who first introduced it in (Schiffer, 1978), but (Recanati, 1993).

set them apart from indexicals is that the descriptive content would not be semantically transparent, i.e., it would eliminate the salience of any other property in the context, making them unsusceptible to quasi-anaphoric interpretation, both expressed or implicated. That would be in accordance with the general rule that descriptive modes of presentations influence the pragmatics of expressions.

8 Conclusion

My analysis of the descriptive uses of indexicals provides an explanation for the pragmatic differences between indexicals and referentially used definite descriptions without forcing the conclusion that descriptions are semantically uniform. This is not an argument for the semantic ambiguity of descriptions, but can be seen as a refutation of attempted arguments against it, such as Nunberg's – arguments which make explicit or implicit use of Neale's Hypothesis.

Katarzyna Kijania-Placek
Institute of Philosophy, Jagiellonian University
52 Grodzka, 31-044 Krakow, Poland
katarzyna.kijania-placek@uj.edu.pl

References

Devitt, M. (2004). The case for referential descriptions. In M. Reimer & A. Bezuidenhout (Eds.), *Descriptions and beyond* (pp. 280–305). Oxford: Clarendon Press.

Donnellan, K. (1966). Reference and definite descriptions. *Philosophical Review*, 75, 281–304.

Grice, H. P. (1989). *Studies in the way of words*. Cambridge, Massachusetts: Harvard University Press.

Kijania-Placek, K. (2009). Anaforyczna interpretacja deskryptywnych użyć wyrażeń okazjonalnych. *Filozofia Nauki*, 1(65), 31–39.

Kijania-Placek, K. (200x). *Anaphoric interpretation of descriptive indexicals*. (Forthcoming)

Kripke, S. (1977). Speaker's reference and semantic reference. In P. A. French, T. Uehling, Jr., & H. K. Wettstein (Eds.), *Midwest studies in philosophy, volume II* (pp. 255–276). Minneapolis: University of Minnesota Press. (Volume 2: Studies in Semantics.)

Linsky, L. (1963). Reference and referents. In C. Caton (Ed.), *Philosophy and ordinary language* (pp. 74–89). Urbana: University of Illinois Press.

Neale, S. (1990). *Descriptions.* Cambridge, Massachusetts: The MIT Press.

Neale, S. (1993). Term limits. In J. E. Tomberlin (Ed.), *Philosophical perspectives, volume 7: Language and logic* (pp. 89–123). Oxford: Blackwell Publishers.

Nunberg, G. (1993). Indexicality and deixis. *Linguistics and Philosophy, 16*(1), 491–538.

Nunberg, G. (2004). Descriptive indexicals and indexical descriptions. In M. Reimer & A. Bezuidenhout (Eds.), (pp. 261–279). Oxford: Clarendon Press.

Recanati, F. (1993). *Direct reference: From language to thought.* Oxford: Blackwell Publishers.

Recanati, F. (2005). Deixis and anaphora. In Z. G. Szabó (Ed.), *Semantics versus pragmatics* (pp. 286–316). Oxford: Oxford University Press.

Reimer, M., & Bezuidenhout, A. (Eds.). (2004). *Descriptions and beyond.* Oxford: Clarendon Press.

Schiffer, S. (1978). The basis of reference. *Erkenntnis, 13,* 171–206.

Wettstein, H. K. (1981). Demonstrative reference and definite descriptions. *Philosophical Studies, 40,* 241–257.

On Relevance Conditions for Asserting Disjunctions

Hans Lycke*

1 Introduction

Communication is a goal–directed activity. In general, it can serve multiple purposes, e.g., information transfer, expression of emotions, making promises,... According to H.P. Grice (1989), if communication is to be successful, the specific purpose of the communicative act has to be known and accepted by all participants. Hence, for a participant to be cooperative (i.e. willing to make the communication successful), she should tailor her contributions to the specific purpose of the communicative act she is involved in (the *cooperative principle*).

> *The Cooperative Principle*
> Make your conversational contribution such as is required, at the stage at which it occurs, by the accepted purpose or direction of the talk exchange in which you are engaged. (Grice, 1989, p. 26)

Grice proposed his infamous maxims (of quantity, quality, relation, and manner) in order to specify the main characteristics of communicative acts governed by the cooperative principle.

Contrary to appearances, the Gricean maxims should not be interpreted as normative statements directed at speakers. As made clear in the neo–Gricean literature — see, e.g., Bach (2006), Horn (2004), and Levinson (2000) —, the maxims should be interpreted more broadly as "presumptions about utterances, presumptions that we as listeners rely on and as speakers exploit". (Bach, 2006, p. 24) Despite the fact that both listeners and speakers are referred to, neo–Griceans usually

---

*The author is a Postdoctoral Fellow of the Special Research Fund of Ghent University.
I would like to thank Curtis Franks, Andreas Pietz, and Greg Restall for their valuable remarks at LOGICA2009 (Hejnice, Czech Republic). Moreover, I'm also indebted to Giuseppe Primiero for helpful comments during the preparation of this paper.

focus on listeners, that is on how they rely on the Gricean maxims to retrieve the intended meaning of an utterance. In this paper though, I will focus on speakers. More specifically, on how they efficiently exploit the Gricean maxims to produce utterances that bring the communicative act they are involved in closer to the fulfillment of its purpose. In other words, I will investigate the Gricean behavior of *cooperative speakers*. Moreover, I will do so from a formal point of view.

Relevance Conditions for Asserting Disjunctions. Because of the vastness of the subject, I will only deal with the relevance conditions related to the disjunction, i.e. the conditions that determine whether it is relevant to assert a disjunctive statement. Whether the approach can be extended to other connectives as well (e.g., conjunction, implication,...) is left for further research.

The relevance conditions related to the disjunction are most easily explained for *atomic disjunctions*, i.e. disjunctions for which the disjuncts are atomic formulas or negations of atomic formulas. Hence, before turning to the relevant assertability of disjunctions (formulas respectively) in general, the simpler case of atomic disjunctions will be discussed first.

For an atomic disjunction $A \vee B$ to be relevantly assertable, three conditions have to be fulfilled. First of all, the speaker obviously has to know that $A \vee B$ is the case. Secondly, neither A nor B may be known by the speaker, otherwise she isn't as informative as she could be. Finally, the speaker has to know whether A and B are co–consistent, which means that she has to know whether $A \wedge B$ is consistent. If the latter is not the case, $A \vee B$ is a tautology, so that one can hardly claim that it is relevant to assert it, for it's informational content is empty.

A Relevantly Assertable Atomic Disjunction
A speaker s may assert an atomic disjunction $A \vee B$ in case

(i) she knows that $A \vee B$ is the case,

(ii) she doesn't know that A is the case,

(iii) she doesn't know that B is the case, and

(iv) she doesn't know that $A \wedge B$ is inconsistent.

The case for atomic disjunctions is generalized to all disjunctions (formulas respectively) by means of the following demands: condition (i) should apply to the disjunction (formula respectively) as a whole, and conditions (ii)–(iv) should apply to all "disjunctive subformulas". Both demands need some explanation. First of all, to see why condition (i) has to apply to the disjunction (formula respectively) in general and not to all its disjunctive subformulas, consider the disjunction $p \vee (q \vee r)$. If the speaker has to know the disjunctive subformula $q \vee r$ in order for $p \vee (q \vee r)$ to be relevantly assertable, then the latter cannot ever be relevantly assertable, for it is impossible to satisfy condition (iii).

Secondly, what is meant by the second demand is actually not as simple as it seems to be. Concerning conditions (ii)–(iii), what is meant is the following: where $A[B \vee C]$ expresses that $B \vee C$ is a (positive) subformula of A, neither $A[B]$ nor $A[C]$ may be known by the speaker in order for A to be relevantly assertable. To see why this should be the case, consider again the formula $p \vee (q \vee r)$. This disjunction isn't relevantly assertable in case either p, $q \vee r$, q or r is known by the speaker (the disjuncts of the disjunctive subformulas of $p \vee (q \vee r)$). However, also in case that either $p \vee q$ or $p \vee r$ is known by the speaker, the formula $p \vee (q \vee r)$ should obviously not be relevantly assertable. Only in case conditions (ii)–(iii) are interpreted as above, this will be the case. Concerning condition (iv), what is meant is the following: for $A[B_1 \vee (C_1[B_2 \vee (C_2[\ldots[B_n \vee C_n]\ldots])])]$ to be relevantly assertable, $B_1, B_2, \ldots, B_n$, and C_n have to be co–consistent. The reasoning is quite similar to the one for conditions (ii)–(iii) and is left to the reader.

The Dynamics of Relevant Assertability. In the neo–Gricean literature, conditions (ii)–(iv) are usually considered to be *implicated* (i.e. derived on non–deductive grounds) by listeners upon hearing the utterance $A \vee B$ — see, e.g., Levinson (2000, p. 19). However, speakers obviously will not implicate them from $A \vee B$, for (ii)–(iv) are *relevance conditions* that determine whether or not $A \vee B$ is relevantly assertable. Hence, knowledge of these statements has to be obtained irrespective of condition (i). Nonetheless, there is a striking correspondence between implicatures and relevance conditions. Both are only derivable *in a defeasible way*! For a discussion of the defeasibility of implicatures, see, e.g., Jaszczolt (2008) and Levinson (2000, ch. 1). That relevance conditions are also defeasible is easily verified. First of all, new information may become available, as a consequence of which some disjunctions might not be relevantly assertable anymore (actually, this comes down to *non–monotonicity*). For example, as long as Mary doesn't know that John is at home (and not at work), she may relevantly assert that John is either at home or at work. However, when she comes to know that John is sick and has gone home, she can't relevantly assert anymore that John is either at work or at home. To be relevant, she will now have to assert that John is at home. Secondly, as people are *not logically omniscient* (they do not know all consequences of their knowledge base), people might also gain a better insight in what they already know. This might also result in some disjunctions not being relevantly assertable anymore. For example, suppose that Mary knows that it is Friday as well as that John doesn't work on Fridays. For as long as she hasn't made the connection between both facts, she will consider it relevant to assert that John is either at work or at home. Once she has made the connection though, i.e., once she has derived from both facts that John won't be at work, she will have to assert that John is at home in order to remain relevant (from a formal point of view, this is a strictly proof theoretic feature, and not a metatheoretic one — see, e.g., (Batens, 2007)).

Aim of this Paper. In this paper, I will provide a formal explication of the way speakers make use of the relevance conditions discussed above in order to determine whether it is relevant to assert a particular disjunction (formula respectively). First of all, I will show that the relevance conditions under consideration can be represented adequately by means of the standard bimodal logic **KC**. Secondly, in order to formally explicate the defeasible nature of some of these relevance conditions, I will make use of the *adaptive logics framework* — see, e.g., Batens (2007) and Batens et al. (ta). More specifically, I will present the adaptive logic **RIT**^{s} (based on the logic **KC**) that will treat the relevance conditions (ii)–(iv) as *defeasible presuppositions*, which means that their truth will be presupposed unless or until this can no longer be done. As a consequence, a particular disjunction (formula respectively) will only be considered as relevantly assertable in a conditional way (for the time being, given the speaker's limited knowledge and limited insight in her knowledge base).

2 Formally Representing Relevance Conditions

In view of the informal discussion above, a formal approach towards the relevance conditions for asserting disjunctions should be able to represent both *knowledge* and *consistency*. In this section, I will present the logic **KC** (Knowledge & Consistency) that is able to do so. The logic **KC** is a standard bimodal logic, obtained by adding two modal operators to (propositional) *classical logic*, viz. the *epistemic* operator K (capturing knowledge) and the *metatheoretic* operator C (capturing consistency).

As in **KC**, the formulas KA and CA are used respectively to express that the formula A is known by the speaker, and to express that the formula A is consistent, the relevance conditions (i)–(iv) for asserting atomic disjunctions (see section 1) are represented as follows:

Formal Representation of Relevantly Assertable Atomic Disjunctions
A speaker s may assert an atomic disjunction $A \vee B$ in case the following four conditions are satisfied:

(i) K$(A \vee B)$,
(ii) ¬KA,
(iii) ¬KB, and
(iv) KC$(A \wedge B)$.

How the modal operators K and C are used to express the relevance conditions for asserting disjunctions (formulas respectively) in general will be explained later on (in section 2.2). First, the logic **KC** will be characterized in some more detail (in section 2.1).

2.1 The Bimodal Logic KC

The bimodal logic **KC** is based on the language $\mathcal{L}^\mathcal{M}$, obtained by adding the epistemic operator K, as well as the metatheoretic operator C, to the standard propositional language $\mathcal{L}$ (see also table 1). The set of well–formed formulas (wffs) $\mathcal{W}^\mathcal{M}$ of $\mathcal{L}^\mathcal{M}$ is defined in the usual way.

Language	Letters	Logical Symbols	Set of Formulas
$\mathcal{L}$	$\mathcal{S}$	$\neg, \wedge, \vee, \supset, \equiv$	$\mathcal{W}$
$\mathcal{L}^{\mathcal{M}}$	$\mathcal{S}$	$\neg, \wedge, \vee, \supset, \equiv, \mathsf{K}, \mathsf{C}$	$\mathcal{W}^{\mathcal{M}}$

Table 1: The Languages $\mathcal{L}$ and $\mathcal{L}^{\mathcal{M}}$.

Two remarks are necessary. First of all, in the remaining of this paper, only negation, disjunction, the epistemic operator K, and the metatheoretic operator C are taken as primitive. The other logical symbols are defined as usual.

Secondly, the epistemic operator K corresponds to the necessity operator of the normal modal logic **S5**, while the consistency operator C corresponds to the possibility operator of the normal modal logic **GL**.[1] Hence, given that they correspond to the modal operators of different modal logics, the operators K and C are not interdefinable. Obviously, their dual operators can easily be defined in **KC**, but since these are of no importance for the remaining of this paper, there is no need to do so.

Proof Theory and Semantics. Syntactically, the logic **KC** is fully characterized by the axiom system of (propositional) classical logic, extended by the axiom schemas and inference rules presented in table 2.

MAx1	$\mathsf{K}(A \supset B) \supset (\mathsf{K}A \supset \mathsf{K}B)$
MAx2	$\mathsf{K}A \supset A$
MAx3	$\mathsf{K}A \supset \mathsf{K}\mathsf{K}A$
MAx4	$A \supset \mathsf{K}\neg\mathsf{K}\neg A$
MAx5	$\neg\mathsf{C}\neg(A \supset B) \supset (\neg\mathsf{C}\neg A \supset \neg\mathsf{C}\neg B)$
MAx6	$\neg\mathsf{C}\neg((\neg\mathsf{C}\neg A) \supset A) \supset \neg\mathsf{C}\neg A$
NEC1	$\vdash A \Rightarrow \vdash \mathsf{K}A$
NEC2	$\vdash A \Rightarrow \vdash \neg\mathsf{C}\neg A$

Table 2: Additional Axiom Schemas and Inference Rules of **KC**.

---

[1] The characterization of the modal operators K and C as **S5**–necessity and **GL**–possibility respectively is completely in accordance with the standard literature on *epistemic logic*, see, e.g., (Hintikka, 2005), and the *logic of provability*, see, e.g., (Boolos, 1993). For an overview of both, see Garson (2006).

Due to space limitations, I will not provide a semantic characterization of the logic **KC**. Nothing fundamental is lost though, for K and C are characterized as in **S5** and **GL**, respectively. Hence, the semantics of **KC** can easily be obtained by any reader acquainted with these logics.

2.2 Formally Representing Relevantly Assertable Disjunctions

Natural language sentences are taken to be represented by means of $\mathcal{L}$–wffs (well–formed formulas of the language $\mathcal{L}$, see table 1). The relevant assertability of these sentences is however expressed by means of $\mathcal{L}^\mathcal{M}$–wffs (well–formed formulas of the language $\mathcal{L}^\mathcal{M}$, see table 1). Hence, to formally express that the natural language sentence represented by the $\mathcal{L}$–wff A (henceforth, the $\mathcal{L}$–wff A) is relevantly assertable, the $\mathcal{L}$–wff A is mapped to the appropriate $\mathcal{L}^\mathcal{M}$–wff A'. This mapping is done by the functions g and g^* that will be characterized below. First however, consider the following preliminary definition.

Definition 1. For $\Delta = \{B_1, \ldots, B_n\}$, $\bigvee(\Delta) =_{df} B_1 \vee \cdots \vee B_n$ and $\bigwedge(\Delta) =_{df} B_1 \wedge \cdots \wedge B_n$

Next, where $\mathcal{W}^\circ$ is the set of all subsets of $\mathcal{W}$ (the set of $\mathcal{L}$–wffs, see table 1), the functions g and g^* are characterized as follows:[2]

G1.0 $g : \mathcal{W} \times \mathcal{W}^\circ \to \mathcal{W}^\mathcal{M}$.

G1.1 For $A \in \mathcal{S}$, $g(A, \Delta) = A$.

G1.2 $g(\neg A, \Delta) = \neg g^*(A, \Delta)$.

G1.3 $g(A \vee B, \Delta) = (g(A, \Delta \cup \{B\}) \vee g(B, \Delta \cup \{A\})) \wedge \neg \mathsf{K} \bigvee(\{A\} \cup \Delta) \wedge \neg \mathsf{K} \bigvee(\{B\} \cup \Delta) \wedge \mathsf{KC} \bigwedge(\{A, B\} \cup \Delta)$.

G2.0 $g^* : \mathcal{W} \times \mathcal{W}^\circ \to \mathcal{W}^\mathcal{M}$.

G2.1 For $A \in \mathcal{S}$, $g^*(A, \Delta) = A$.

G2.2 $g^*(\neg A, \Delta) = \neg g(A, \Delta)$.

G2.3 $g^*(A \vee B, \Delta) = g^*(A, \Delta) \vee g^*(B, \Delta)$.

---

[2] Remember that conjunction, implication, and equivalence are treated as defined connectives. Hence, they are not considered here.

Finally, definition 2 below lays down when it is relevant for a speaker to assert a particular $\mathcal{L}$–wff A.

Definition 2 (Formal Representation). The $\mathcal{L}$–wff A is relevantly assertable by a speaker s iff s knows the $\mathcal{L}^{\mathcal{M}}$–wff $g(A, \emptyset)$ iff $\mathsf{K}g(A, \emptyset)$.

Given the characterization of g and g^* above, it is easily verified that definition 2 nicely incorporates the relevance conditions for asserting disjunctions (as set out in section 1). For, the above characterization (and in particular G1.3) ensures that when the speaker knows the formula $g(A, \emptyset)$, she then knows the formula A, doesn't know any of the disjuncts of the "disjunctive subformulas" of A, and knows the co–consistency of all disjuncts of the "disjunctive subformulas" of A.

Formal Explication of Gricean Behavior. Although the logic **KC** is clearly powerful enough to express the relevance conditions for asserting disjunctions, **KC** isn't strong enough to determine the formulas that are relevantly assertable by a speaker s in view of her knowledge base Γ^{K}. For, in order to be strong enough, the logic **KC** should be able to derive all relevantly assertable formulas $\mathsf{K}g(A, \emptyset)$ from Γ^{K}. Given that the knowledge base Γ^{K} is defined as the set of all $\mathcal{L}$–wffs the speaker s knows to be true (see definition 3), **KC** obviously isn't able to do so (as none of the relevance conditions (ii)–(iv) can be derived from a knowledge base by means of **KC**).

Definition 3. Γ^{K} is the knowledge base of the speaker s iff $\Gamma^{\mathsf{K}} = \{\mathsf{K}A \mid A \in \mathcal{W} \text{ and } s \text{ knows } A\}$.

Because the logic **KC** doesn't enable one to derive enough consequences from the knowledge base of a speaker s, a different logic is used to determine the formulas that are relevantly assertable by s, viz. the logic **RIT^{s}**.

Definition 4 (Gricean Behavior). The $\mathcal{L}$–wff A is relevantly assertable by a speaker s with knowledge base Γ^{K} iff $\Gamma^{\mathsf{K}} \vdash_{\mathbf{RIT^s}} \mathsf{K}g(A, \emptyset)$.

The logic **RIT^{s}** extends the logic **KC** by treating the relevance conditions (ii)–(iv) as *defeasible presuppositions*, which means that the relevance conditions are presupposed to be true *unless* or *until* this can no longer be done. In this way, the logic **RIT^{s}** does not only enable one to determine all formulas that are relevantly assertable by a speaker in view of her knowledge base, the logic **RIT^{s}** also captures the specific dynamics related to the Gricean behavior of cooperative speakers (as described in section 1).

3 The Adaptive Logic RIT^s

The logic **RIT^s** is a standard adaptive logic. Hence, it is characterized by three elements: a *lower limit logic* (**LLL**), a *set of abnormalities* Ω (a set of formulas characterized by a logical form F), and an *adaptive strategy*. Before I will characterize these elements for the logic **RIT^s**, I will first show how they interact in general. This will be done by presenting a short, rather intuitive characterization of the *standard format* of adaptive logics — for an extensive characterization, see Batens (2007) and Batens et al. (ta).

3.1 The Standard Format

Adaptive logics (**AL**) are a branch of logics that was developed to characterize inference relations that lack a positive test,[3] such as e.g. inconsistency–handling, induction, abduction,..., see, e.g., (Batens, 2007). To capture such inference relations, **AL** display a twofold dynamics, an external and an internal one. The former comes down to non–monotonicity (enlarging a premise set Γ may necessitate the withdrawal of consequences derivable from Γ alone), while the latter is a strictly proof–theoretic feature (deriving new consequences from a premise set Γ may necessitate the withdrawal or rehabilitation of earlier derived or withdrawn consequences of Γ). This twofold dynamics is the result of the specific interplay between the constituting elements of an adaptive logic, viz. its **LLL**, its set of abnormalities, and its adaptive strategy.

The **LLL** is the stable part of an adaptive logic. As theorem 1 clearly shows, this means that all **LLL**–consequences of a premise set are also **AL**–consequences of that premise set.[4]

Theorem 1. $Cn_{\mathbf{LLL}}(\Gamma) \subseteq Cn_{\mathbf{AL}}(\Gamma)$.

An adaptive logic typically enables one to derive more consequences from a premise set than its **LLL**. The supplementary **AL**–consequences are obtained by interpreting a premise set *as normally*

---
[3] Phenomena for which there are no finite means to determine whether a formula belongs to the consequence set of a particular premise set.

[4] For a proof of theorem 1, see Batens (2007, p. 237).

as possible. This is done by interpreting as false as much elements of Ω (abnormalities) as possible, which comes down to the fact that some formulas are *conditionally derivable* from a premise set: if $\Gamma \vdash_{\mathbf{LLL}} A \vee Dab(\Delta)$, with $Dab(\Delta)$ a Dab–formula (a finite disjunction of abnormalities), then the formula A is an **AL**–consequence of Γ *unless* or *until* there are reasons to consider some elements of Δ as true. Hence, at this point, the formula A might be called a *conditional consequence* of Γ.

Definition 5. A is a conditional consequence of Γ iff there is some finite $\Delta \subset \Omega$ such that $\Gamma \vdash_{\mathbf{LLL}} A \vee Dab(\Delta)$.

Which of the conditional consequences of a premise set Γ are also *final* consequences of Γ, i.e., **AL**–consequences of Γ, depends on the Dab–consequences of Γ as well as on the adaptive strategy of the adaptive logic. The Dab–consequences are those Dab–formulas that are **LLL**–derivable from Γ.

Definition 6. $Dab(\Delta)$ is a Dab–consequence of Γ iff $\Gamma \vdash_{\mathbf{LLL}} Dab(\Delta)$.

Not all abnormalities occurring in a Dab–consequence can be considered as false. Otherwise, the Dab–consequence itself cannot possibly be considered as true (**LLL**–derivable respectively). Hence, if a premise set has Dab–consequences, some of the conditional consequences of that premise set have to be rejected (because they were derived by mistakenly interpreting some of the abnormalities in a Dab–consequence as false). Which of the conditional consequences will be rejected, is determined by the adaptive strategy. For, the latter provides the guidelines of how to cope with the abnormalities occurring in the Dab–consequences of a premise set. Obviously, different adaptive strategies provide different guidelines. Hence, different strategies will yield different consequence sets (for an overview of the most common adaptive strategies, see (Batens et al., ta)).

It is now easy to see why **AL** display the twofold dynamics mentioned earlier on. First, consider non–monotonicity. Enlarging a premise set Γ may lead to the derivation of Dab–consequences that are not derivable from Γ alone. As a consequence, some of the final consequences of Γ may not be final consequences of the enlarged premise set. Secondly, consider the internal (proof theoretic) dynamics. At a certain stage of an **AL**–proof, a formula A may be considered a consequence of a premise set Γ. For, given the formulas derived at that stage, there might be no reason to presuppose otherwise (for example, because no Dab–consequences have been derived yet). However, at some later stage this might change, for example in case a Dab–consequence has been derived necessitating the withdrawal of A as a consequence of Γ.

3.2 General Characterization of the Logic RIT^s

As the adaptive logic **RIT^s** is an extension of the logic **KC**, the latter is the **LLL** of the former. Consequently, all **KC**–consequences of a premise set are also **RIT^s**–consequences of that premise set.

Theorem 2. $Cn_{\mathbf{KC}}(\Gamma) \subseteq Cn_{\mathbf{RIT^s}}(\Gamma)$.

Next, the set of abnormalities Ω of the logic **RIT^s** is defined as the union of the following two sets:

Definition 7. $\Omega_1 = \{\mathsf{K}A \mid A \in \mathcal{W}\}$.

Definition 8. $\Omega_2 = \{\neg\mathsf{KC}(A \wedge B) \mid A, B \in \mathcal{W}\}$.

To interpret a premise set as normally as possible, the logic **RIT^s** will presuppose as many elements of Ω as possible as false. Hence, given the validity in **KC** of the law of excluded middle, $\neg\mathsf{K}A$ and $\mathsf{KC}(A \wedge B)$ will be conditional consequences of any premise set Γ (for any $A, B \in \mathcal{W}$). This indeed means (as claimed before) that the logic **RIT^s** treats the relevance conditions for asserting disjunctions as defeasible presuppositions.

Finally, the adaptive strategy of **RIT^s** is the *normal selections* strategy. In general, this means that a conditional consequence of a premise set Γ, derived by interpreting the Dab–formula $Dab(\Delta)$ as false, is an **AL**–consequence of Γ only in case $Dab(\Delta)$ is not **LLL**–derivable from Γ. In view of definition 5, this implies that **RIT^s**–derivability is defined as follows:

Definition 9. $\Gamma \vdash_{\mathbf{RIT^s}} A$ iff there is some finite $\Delta \subset \Omega$ such that $\Gamma \vdash_{\mathbf{KC}} A \vee Dab(\Delta)$ (A is a conditional consequence of Γ) and $\Gamma \nvdash_{\mathbf{KC}} Dab(\Delta)$.

Examples. Because of space limitations, no semantic or proof theoretic characterization of the logic **RIT^s** will be provided. Nonetheless, based on the general characterization above, some simple examples will be given to clarify the way in which the logic **RIT^s** captures the Gricean behavior of cooperative speakers. First, consider an example of a disjunction that is not relevantly assertable by the speaker because she knows one of the disjuncts.

Example 1. Let the knowledge base Γ^K of a speaker s be the set $\{\mathsf{K}(p\vee q), \mathsf{K}p\}$. For $p\vee q$ to be relevantly assertable by s, $\mathsf{K}g(p\vee q, \emptyset)$ has to be derivable from Γ^K by means of the logic **RIT^s**. As $\Gamma^K \vdash_{\mathbf{KC}} \mathsf{K}g(p\vee q, \emptyset) \vee \mathsf{K}p \vee \mathsf{K}q \vee \neg\mathsf{KC}(p\wedge q)$, $\mathsf{K}g(p\vee q, \emptyset)$ is a conditional consequence of Γ^K. However, $\Gamma^K \vdash_{\mathbf{KC}} \mathsf{K}p$, which means that $\mathsf{K}g(p \vee q, \emptyset)$ is not a final consequence of Γ^K. Hence, $p \vee q$ is not relevantly assertable by the speaker s. On the other hand, it is easily verified that p is relevantly assertable by s.

Secondly, consider an example of a disjunction that is not relevantly assertable because the informational content of one of its "disjunctive subformulas" is empty.

Example 2. Let the knowledge base Γ^K of a speaker s be the set $\{\mathsf{K}p\}$. For $(p\wedge q)\vee\neg q$ to be relevantly assertable by s, $\mathsf{K}g((p\wedge q)\vee\neg q, \emptyset)$ has to be derivable from Γ^K by means of the logic **RIT^s**. As $\Gamma^K \vdash_{\mathbf{KC}} \mathsf{K}g((p\wedge q)\vee\neg q, \emptyset) \vee \mathsf{K}(p\wedge q) \vee \mathsf{K}\neg q \vee \neg\mathsf{KC}((p\wedge q)\wedge\neg q)$, $\mathsf{K}g((p\wedge q)\vee\neg q, \emptyset)$ is a conditional consequence of Γ^K. However, $\Gamma^K \vdash_{\mathbf{KC}} \neg\mathsf{KC}((p\wedge q)\wedge\neg q)$, which means that $\mathsf{K}g((p\wedge q) \vee \neg q, \emptyset)$ is not a final consequence of Γ^K. Hence, $(p \wedge q) \vee \neg q$ is not relevantly assertable by the speaker s. On the other hand, it is easily verified that p is relevantly assertable by s.

4 Conclusion

In this paper, I have initiated the formal explication of the Gricean behavior of cooperative speakers (i.e. how cooperative speakers treat the conditions that guarantee the relevant assertability of particular formulas). I have done so by presenting the adaptive logic **RIT^s** that captures the way cooperative speakers handle the relevance conditions related to the assertion of disjunctions.

Hans Lycke
Centre for Logic and Philosophy of Science, Ghent University
Blandijnberg 2, 9000 Gent, Belgium

Hans.Lycke@UGent.be
http://logica.ugent.be/hans/

References

Bach, K. (2006). The top ten misconceptions about implicature. In B. Birner & G. Ward (Eds.), *Drawing the boundaries of meaning: Neo-Gricean studies in pragmatics and semantics in honor of Laurence R. Horn* (pp. 21–30). Amsterdam: John Benjamins.

Batens, D. (2007). A universal logic approach to adaptive logics. *Logica Universalis*, *1*, 221–242.

Batens, D., Meheus, J., & Provijn, D. (ta). An adaptive characterization of signed systems for paraconsistent reasoning.

Boolos, G. (1993). *The logic of provability*. Cambridge (UK): Cambridge University Press.

Garson, J. (2006). *Modal logic for philosophers*. Cambridge (UK): Cambridge University Press.

Grice, H. (1989). *Studies in the way of words*. Cambridge (Mass.): Harvard University Press.

Hintikka, J. (2005). *Knowledge and belief: An introduction to the logic of the two notions*. London: King's College. (*Texts in philosophy*, vol. 1, prepared by V. Hendricks & J. Symons)

Horn, L. R. (2004). Implicature. In L. R. Horn & G. Ward (Eds.), *Handbook of pragmatics* (pp. 3–28). Oxford: Blackwell Publishing.

Jaszczolt, K. M. (2008, Fall). Defaults in semantics and pragmatics. In E. N. Zalta (Ed.), *The stanford encyclopedia of philosophy*. Available from http://plato.stanford.edu/archives/fall2008/entries/defaults-semantics-pragmatics/

Levinson, S. C. (2000). *Presumptive meanings. The theory of generalized conversational implicature*. Cambridge (Mass.): MIT Press.

Logic of Questions from the Viewpoint of Dynamic Epistemic Logic

Michal Peliš and Ondrej Majer[*]

1 Introduction

Questions play an important role in communication in both natural language and artificial languages. However, the logic of questions — the discipline of which the aim is to represent questions as logical entities and to explore relations between these entities — has been receiving relatively little attention in literature until quite recently.

The logic of questions has, maybe surprisingly, a long history (cf. (Harrah, 2002)). F. Cohen and R. Carnap seem to be the first authors attempting to formalize questions in a logical framework — their attempts date back to the 1920s. The first 'boom' of logical approach to questions took place in the 1950s (Hamblin, Prior, Stahl) and continued in the 1960s (Åqvist, Hintikka, Kubiński). The first comprehensive monograph on questions was published in the 1970s (Belnap & Steel, 1976). The 1990's gave birth to several approaches, some of which are still influential. The most important are the semantic approach of Jeroen Groenendijk and Martin Stokhof and Inferential Erotetic Logic of Andrzej Wiśniewski. Recently the logic of questions receives more attention in connection with dynamic logic and game theory (e.g., (Benthem & Minică, 2009) and (Genot, 2009)).

Our main goal is to consider questions in the process of communication. When an agent asks a question, she does not only provide some facts to her listeners, but she also reveals the structure of her knowledge, in particular, what she presupposes and what she considers as possible updates of her current knowledge. We see asking a question as a process of information exchange, which can be quite

---

[*]Work on this paper was supported in part by grant no. 401/07/0904 of the Grant Agency of the Czech Republic and in part by grant no. IAA900090703 of the Grant Agency of the Academy of Sciences of the Czech Republic. We wish to thank to Nuel Belnap, Jaroslav Peregrin, Martina Pivoňková, and Andrzej Wiśniewski for valuable comments.

complex. From a logician's point of view the most appropriate framework for representing this process is the one of dynamic epistemic logic. The machinery of this framework allows for the representation of information states of individual agents and their changes in the process of receiving information from the other agents. We focus on the case when new information is 'public', i.e., available to all agents in a particular group, typically the new information is announced so that everybody in the group can hear it. This very situation serves also as a motivation of the public announcement logic (see (Ditmarsch, Hoek, & Kooi, 2008)), which we shall employ.

2 Public announcement logic

Public announcement logic in the sense of (Ditmarsch et al., 2008) is an extension of the multi-agent epistemic logic S5, which is a multimodal logic, where each of the finitely many modal operators is an S5-modality.

We start with a propositional multimodal language $\mathcal{L}^K$ — a classical propositional language $\mathcal{L}$ with finitely many modalities K_i to be read as 'the agent i knows …'. The formulas of $\mathcal{L}^K$ are defined in a standard way ($p, q, \ldots$ are signs for atomic formulas):

$$\varphi ::= p \mid \neg \psi \mid \psi_1 \vee \psi_2 \mid \psi_1 \wedge \psi_2 \mid \psi_1 \rightarrow \psi_2 \mid \psi_1 \leftrightarrow \psi_2 \mid K_i \psi.$$

As in the standard modal logics we can define a dual operator to K_i — an epistemic possibility M_i ('the agent i admits that possibly…'): $M_i \varphi \equiv \neg K_i \neg \varphi$.

The semantics of $\mathcal{L}^K$ is based on the standard Kripke-style models. A *Kripke frame* is a relational structure $\mathcal{F} = \langle S, R_1, \ldots, R_m \rangle$ with a set of states (points, indices, possible worlds) S and accessibility relations $R_i \subseteq S^2$, for agents $i \in \{1, \ldots, m\}$. *Kripke model* $\mathbf{M}$ is a pair $\langle \mathcal{F}, v \rangle$ where v is a valuation of atomic formulas. The satisfaction relation $\models$ is defined in a standard way:

- $(\mathbf{M}, s) \models p$ iff $(\mathbf{M}, s) \in v(p)$,
- $(\mathbf{M}, s) \models \neg \varphi$ iff $(\mathbf{M}, s) \not\models \varphi$,
- $(\mathbf{M}, s) \models \psi_1 \vee \psi_2$ iff $(\mathbf{M}, s) \models \psi_1$ or $(\mathbf{M}, s) \models \psi_2$,
- $(\mathbf{M}, s) \models \psi_1 \wedge \psi_2$ iff $(\mathbf{M}, s) \models \psi_1$ and $(\mathbf{M}, s) \models \psi_2$,

- $(\mathbf{M}, s) \models \psi_1 \to \psi_2$ iff $(\mathbf{M}, s) \models \psi_1$ implies $(\mathbf{M}, s) \models \psi_2$,
- $(\mathbf{M}, s) \models K_i \varphi$ iff $(\mathbf{M}, s_1) \models \varphi$, for each s_1 such that $s R_i s_1$.

Each R_i is an equivalence relation on S^2.

On top of the individual epistemic modalities K_i and M_i we introduce group modalities E_G and C_G. $E_G \varphi$ means 'each agent from $G \subseteq \{1, \ldots, m\}$ knows φ', i.e.,

$$E_G \varphi \leftrightarrow \bigwedge_{i \in G} K_i \varphi.$$

We shall call E_G *group knowledge*. Let us stress, that E_G does not guarantee that a member of the group G knows that she shares the same information with some other members of the group.

The second group modality C_G is stronger, $C_G \varphi$ requires not only that φ is group knowledge, but also that this fact is reflected by everybody in the group G (everybody knows that everybody knows φ and everybody knows that everybody knows that everybody knows φ, etc.). We can see C_G as an infinite conjunction of all finite iterations of the group knowledge E_G:

$$C_G \varphi \leftrightarrow \varphi \wedge E_G \varphi \wedge E_G E_G \varphi \wedge E_G E_G E_G \varphi \wedge \ldots$$

This knowledge, called *common knowledge*, is maximally shared in the sense that everybody from G is aware of it. Such type of knowledge is essential for collective behavior and coordination of collective actions.

As we said E_G is definable in the language $\mathcal{L}^K$, so adding group knowledge is just a conservative extension of the background multimodal epistemic logic. However, this is not the case of common knowledge. Multimodal epistemic logic with common knowledge is a stronger system $\mathcal{L}^{KC}$ which we obtain from $\mathcal{L}^K$ adding the operator C_G and the clause:

- $(\mathbf{M}, s) \models C_G \varphi$ iff $(\mathbf{M}, s_1) \models \varphi$ for each s_1 such that $s \left(\bigcup_{i \in G} R_i \right)^* s_1$

$\left(\bigcup_{i \in G} R_i \right)^*$ is a reflexive and transitive closure of $\bigcup_{i \in G} R_i$ and it means that s_1 is accessible from s by each R_i ($i \in G$) in k steps, for any $k \geq 0$. Common knowledge is stronger than group knowledge:

$$C_G \phi \to E_G \phi \to K_i \phi$$

is valid for each $i \in G$.

2.1 Public announcement

Let us imagine a group of three players: Ann, Bill, and Catherine. Each of them has one card and nobody can see the cards of the others. One of the cards is the Joker and everybody knows this fact. Ann received the Joker, but neither Bill nor Catherine know which of the other two players, has it. In particular, both of them are not able to distinguish between the states where Ann has the Joker and where she has not. If Ann publicly announces

"I've got the Joker.",

everybody in the group learns this fact. Situations where Ann does not have the Joker are excluded from the (epistemic) models of Bill and Catherine.

Our example gives a typical situation represented in the public announcement logic — after a public announcement of a statement ϕ ("I've got the Joker"), some other statement ψ holds (e.g., "Bill knows Ann has the Joker and Catherine knows Ann has the Joker"). In fact the author of an announced statement is irrelevant in our framework. The statement is understood as information coming to each member of a group in the same way. From this point of view Ann's announcement in our example has the same effect as if an external observer announces "Ann has the Joker".

Formally we introduce the logic of public announcement as an extension of the system $\mathcal{L}^{KC}$. We define a box-like operator [ ], such that the intended meaning of $[\phi]\psi$ is 'after the public announcement of ϕ, it holds that ψ'. The semantics of the new announcement operator is:

- $(\mathbf{M}, s) \models [\phi]\psi$ iff $(\mathbf{M}, s) \models \phi$ implies $(\mathbf{M}|_\phi, s) \models \psi$

where $\mathbf{M}|_\phi = \langle\langle S', R'_1, \ldots, R'_m\rangle, v'\rangle$ is defined as follows:

$$S' = \{s \in S \mid s \models \phi\},$$
$$R'_i = R_i \cap S'^2,$$
$$v'(p) = v(p) \cap S'.$$

The model $\mathbf{M}|_\phi$ is obtained from $\mathbf{M}$ by deleting of all states where ϕ is not true and by the corresponding restrictions of accessibility relations and the valuation function. Again we can introduce a dual operator

Logic of Questions

$\langle \ \rangle$ defined in a standard way as $\langle \phi \rangle \psi$ iff $\neg[\phi]\neg\psi$. If we rewrite the corresponding semantic clause, we obtain

- $(\mathbf{M}, s) \models \langle \phi \rangle P$ iff $(\mathbf{M}, s) \models \phi$ and $(\mathbf{M}|_\phi, s) \models \psi$.

The intended meaning of the dual operator is 'after a *truthful* announcement of ϕ, it holds that ψ'. It is easy to see that the diamond-like operator is stronger:

Lemma 1. $\models \langle \phi \rangle \psi \to [\phi]\psi$.

The following proposition shows that languages $\mathcal{L}^{K[]}$ and $\mathcal{L}^K$ have the same expressive power — it provides a reduction of formulas with the public announcement operator to the epistemic ones. The corresponding equivalences in the proposition give in fact an axiomatization of the announcement operator in the public announcement epistemic logic without common knowledge (Ditmarsch et al., 2008, p. 81).

Proposition 1. *The following equivalences are valid* ($\circ \in \{\wedge, \vee, \to\}$):

$$[\phi]p \leftrightarrow (\phi \to p),$$
$$[\phi]\neg\psi \leftrightarrow (\phi \to \neg[\phi]\psi),$$
$$[\phi](\psi \circ \chi) \leftrightarrow ([\phi]\psi \circ [\phi]\chi),$$
$$[\phi]K_i\psi \leftrightarrow (\phi \to K_i[\phi]\psi),$$
$$[\phi][\psi]\chi \leftrightarrow [\phi \wedge [\phi]\psi]\chi.$$

For common knowledge there is no such reduction, the language $\mathcal{L}^{KC[]}$ is more expressive than $\mathcal{L}^{KC}$. We only have a rule describing the relationship between the public announcement and common knowledge, e.g., the one in (Ditmarsch et al., 2008, p. 83):

$$\frac{(\chi \wedge \phi) \to [\phi]\psi \wedge E_G \chi}{(\chi \wedge \phi) \to [\phi]C_G \psi}. \tag{1}$$

2.2 Updates

Let us return to our example. As we said, members of a group learn what was announced. In particular, if Ann says

"I've got the Joker.",

the announced fact becomes commonly known in the group of players. This seems to suggest that a publicly announced proposition becomes common knowledge . But what if Ann says:

"You don't know it yet, but I've got the Joker."

Although the formula $(J_a \land \neg K_b(J_a) \land \neg K_c(J_a))$ is true in the moment of announcement, it is evident that its epistemic part ("you don't know it yet") becomes invalid after it is announced. So the formula $(J_a \land \neg K_b(J_a) \land \neg K_c(J_a))$ becomes false after the announcement. Formulas, which became false after they are truthfully announced (as in our example), are in public announcement logic called an *unsuccessful update*; if they are true, we call them a *successful update*. For the [ ] modality we have similar notions. If a formula $[\phi]\phi$ is valid, we call it a *successful formula*, otherwise it is an *unsuccessful formula*.

Definition 1.

- Formula ϕ is a *successful update* in $(\mathbf{M}, s)$ iff $(\mathbf{M}, s) \models \langle \phi \rangle \phi$.

- Formula ϕ is an *unsuccessful update* in $(\mathbf{M}, s)$ iff $(\mathbf{M}, s) \models \langle \phi \rangle \neg \phi$.

- Formula ϕ is a *successful formula* iff $[\phi]\phi$ is valid, otherwise it is an *unsuccessful formula*

If a formula is an unsuccessful update, it cannot be commonly known in the updated model. Using the soundness proof of the rule (1) we can prove that a formula is true after a public announcement if and only if it gets common knowledge after the announcement (see (Ditmarsch et al., 2008, p. 83 and 86)).

Proposition 2. $[\phi]\psi$ *is valid iff* $[\phi]C_G\psi$ *is valid.*

As a consequence we get

Proposition 3. $[\phi]\phi$ *is valid iff* $[\phi]C_G\phi$ *is valid.*

Thus, publicly announced successful formulas are commonly known.[1]

---

[1] Atoms, $K_i\phi$, and $\neg K_i\phi$ (for every ϕ) are examples of successful formulas.

3 Questions

Let us return to our card players. It seems reasonable for Catherine to say:

"Who has got the Joker?"

We recognize this sentence as an *interrogative sentence* because of its word order and the question mark. In speech, we recognize uttering a question because of the intonation and interrogative pronunciation. The *interrogative speech act* is important in the pragmatic approach to questions, where we focus on the fact that a questioner expresses a request to an addressee.

In logic of questions it is generally accepted that interrogative sentences have a different meaning than indicative ones.[2] In our approach to questions we pay special attention to their semantic content. The traditional starting point in the semantics of questions are Hamblin's postulates:

1. An answer to a question is a statement.

2. Knowing what counts as an answer is equivalent to knowing the question.

3. The possible answers to a question are an exhaustive set of mutually exclusive possibilities.

These postulates have been subjected to some criticism, see (Harrah, 2002) and (Groenendijk & Stokhof, 1997). We are not going to join these critical discussions. In our project we consider 'Hamblin's picture' as partially adequate to the formalization of questions in the epistemic framework. The first two postulates seem to be the first step towards the formalization of questions known as *set-of-answers methodology* (SAM, for short). We implement them in our framework using a modification of Wiśniewski's SAM (see, e.g., (Wiśniewski, 1995, 2001) and (Peliš, 2008)) where questions are represented as finite sets of formulas.[3] We shall not in general accept the third postulate,

---

[2] There are exceptions, however, some logicians, e.g., Pavel Tichý hold the view that the logical form of both interrogatives and indicatives are the same, what is different is the pragmatic aspect (Tichý, 1978).

[3] Unlike Wiśniewski, we allow the mixing of declarative and interrogative formulas together, so we can have questions about questions as well.

we shall just use it for delineating a special subclass of questions in section 3.1.

We extend the language of epistemic logic $\mathcal{L}^{KC}$ adding brackets $\{,\}$ and the question mark $?_i$ for a question of an agent i. We obtain the language $\mathcal{L}_Q^{KC}$. A question is any formula of the form

$$?_i\{\alpha_1,\ldots,\alpha_n\},$$

such that $\alpha_1,\ldots,\alpha_n$ are syntactically distinct formulas of our extended epistemic language $\mathcal{L}_Q^{KC}$ and $n \geq 2$ (there are at least two answers). We shall denote a question of an agent i as $Q^i, Q_1^i, \ldots$, etc. As an index we can also use a subset of the set of agents. The intended reading of a question Q^i is:

> 'Is it the case that α_1 or is it the case that α_2 ... or is it the case that α_n?'

We shall call $\{\alpha_1,\ldots,\alpha_n\}$ the set of *direct answers* to the question Q^i and denote it dQ^i.

Let us assume that Catherine from our example asks the question:

> "Who has got the Joker: Bill, or Ann?"

This question has the following direct answers:

- Bill has got the Joker.
- Ann has got the Joker.

Catherine expresses that she

1. does not know any of the answers to the question,
2. considers all the answers to be possible, and, moreover,
3. presupposes what is implicitly included in the answers, i.e., it must be the case that either Bill has got the Joker, or Ann has got it.

Catherine (the agent-questioner) provides the information of her ignorance (item 1) as well as the expected way to extend her knowledge; 2 says that there are two possibilities to make this extension and 3 that exactly one of the possibilities is expected to be true. A publicly asked question allows listeners to male a picture of the questioner's knowledge structure, which is an important part of communication and problem solving in groups.

In semantics for a majority logical systems we speak about the truth/falsity of a formula (in a particular state of a particular model). It is clear that it makes little sense to speak about the truth/falsity of a question. We introduce instead a concept of *askability* of a question; an askable question is a question which is in some sense 'reasonable' to ask in a certain situation. 'Reasonability' corresponds to the following three conditions:

1. **Non-triviality**   It is not reasonable to ask a question if the answer is known.

2. **Admissibility**   Each direct answer is considered as possible.

3. **Context**   At least one of the direct answers must be the right one.

These askability conditions will play the role of the truth conditions in the formal definition of semantics of interrogative formulas.

Definition 2. It holds for a question $Q^i = ?_i\{\alpha_1, \ldots, \alpha_n\}$ that $(\mathbf{M}, s) \models Q^i$ iff

1. $(\forall \alpha \in dQ^i)((\mathbf{M}, s) \not\models K_i \alpha)$
2. $(\forall \alpha \in dQ^i)((\mathbf{M}, s) \models M_i \alpha)$
3. $(\mathbf{M}, s) \models K_i \left(\bigvee_{\alpha_j \in dQ^i} \alpha_j \right)$

We say that Q^i is *askable* in the state $(\mathbf{M}, s)$ (by an agent i).

As we can see, the freedom in the syntactical form of questions was compensated by restrictions in their semantics. We say that a question is (generally) *askable* iff there is a model and a state where the question is askable (by an agent). Askable questions include neither contradiction nor tautology among their direct answers. The former is excluded by the second condition and the latter by the first one. If we work in systems extending classical logic, the first condition is equal to $(\mathbf{M}, s) \models \neg K_i \alpha$, i.e., $(\mathbf{M}, s) \models M_i \neg \alpha$, for each $\alpha \in dQ^i$. We can see the questioner as admitting the possibility of $\neg \alpha$ for each direct answer α to a question Q^i.

Catherine's question

"Who has got the Joker: Bill, or Ann?"

can be formalized by $?_c\{J_b, J_a\}$. This question is askable in a state s whenever there are at least two indistinguishable states with J_b true in one of them and J_a true in the other.

$$\boxed{\begin{array}{ccc} & c & \\ \boxed{x} & \longleftrightarrow & \boxed{y} \\ J_b & & J_a \end{array}}$$

Non-triviality condition says that Catherine is not able to distinguish between the states x and y, both formulas J_b and J_a are considered as possible (admissibility condition), and exactly one of the statements J_b and J_a must be the case (context condition). The shape of our model is influenced by more restrictions given by the context of our card-players example, i.e, commonly known rules of the game (in particular, there is just one Joker in the game).

3.1 Askability, answerhood, and safe questions

Askability is a complex term. We shall discuss now what happens if we violate the askability conditions. Whenever the non-triviality is violated, an agent knows at least one of the direct answers. Then she is able to answer a question. We say that the question is *answered* for her in a state. The violation of the admissibility condition brings about the situation where an agent does not consider at least one of the direct answers possible. She does not need to know any direct answer to a question, but she can reject some of them. Then we say that the question is *partially answered* for her in a state. The violation of the last condition (context) is of another kind. An agent is not aware of what is presupposed. Such a question is called *weakly presupposed* by the agent in a state. Let us sum up the introduced terms:

Definition 3. Let Q^i be a question $?_i\{\alpha_1, \ldots, \alpha_n\}$.

- Q^i is *answered* in $(\mathbf{M}, s)$ (for an agent i) iff $(\mathbf{M}, s) \models \bigvee_{\alpha_j \in dQ^i}(K_i \alpha_j)$. We write $(\mathbf{M}, s) \models A_i Q$.

- Q^i is *partially answered* in $(\mathbf{M}, s)$ (for an agent i) iff $(\mathbf{M}, s) \models \bigvee_{\alpha_j \in dQ^i}(K_i \neg \alpha_j)$. We write $(\mathbf{M}, s) \models P_i Q$.

- Q^i is *weakly presupposed* in $(\mathbf{M}, s)$ (by an agent i) iff $(\mathbf{M}, s) \not\models K_i(\bigvee_{\alpha_j \in dQ^i} \alpha_j)$.

The violation of the context condition means that the agent does not believe that at least one of $\alpha \in dQ$ is the right answer. We shall not discuss this condition as it is not essential for the purposes of this paper. We can avoid the context condition if we restrict ourselves to a special class of questions — *safe questions*, for which it is trivially satisfied. A question Q^i is *safe* if and only if $\left(\bigvee_{\alpha_j \in dQ^i} \alpha_j\right)$ is tautology. This class includes, among others, the well known yes-no questions, i.e., questions of the form $?_i\{\alpha, \neg\alpha\}$.[4]

Definition 4. A question Q^i is (for an agent i)

- *safe in a state* $(\mathbf{M}, s)$ iff $(\mathbf{M}, s_1) \models \left(\bigvee_{\alpha_j \in dQ^i} \alpha_j\right)$, for each s_1 such that $sR_i s_1$.

- *safe* iff $\left(\bigvee_{\alpha_j \in dQ^i} \alpha_j\right)$ is valid.

Another way around the context condition is to define explicitly a common context, which is shared by all members of the group. This is the case of our motivational example — the commonly known context enables us to see Catherine's question $?_c\{J_b, J_a\}$ as safe in the corresponding state.

Catherine's question is a question with mutually exclusive direct answers. We shall make use of a weaker condition — the one where a direct answer does not exclude any of the other direct answers but just some of them. We shall call it *questions with pairs of mutually exclusive direct answers*.[5] There is one more reason to restrict ourselves to the class of questions we just introduced. In general, it is not true that an answered question is also partially answered. $A_i Q$ implies $P_i Q$ only for questions with pairs of mutually exclusive direct answers.

The following proposition is proved in (Peliš, 2009).

Proposition 4. *Let Q^i be a safe question in $(\mathbf{M}, s)$ with pairs of mutually exclusive direct answers. Then the following conditions are equivalent:*

---

[4]The term *safe question* originates from Belnap, the presented meaning is inspired by Wiśniewski.

[5]For example askable safe questions are of this kind. We require askability to avoid safe questions with tautological direct answers.

- $(\mathbf{M}, s) \models \neg Q^i$

- $(\mathbf{M}, s) \models P_i Q$

- There is a formula ϕ such that $(\mathbf{M}, s) \models A_i?\{\phi, \neg\phi\}$ and $Q^i \to ?_i\{\phi, \neg\phi\}$ is valid.

Partial answerhood of a question Q^i in some state is equivalent to the existence of a yes-no question, which is answered at that state and implied by Q^i. From the validity of $Q^i \to ?_i\{\phi, \neg\phi\}$ we know that inaskability of $?_i\{\phi, \neg\phi\}$[6] implies inaskability of Q^i and, therefore, ϕ (as well as $\neg\phi$) imply either some $\alpha \in dQ^i$ or $\neg\alpha$ (for $\alpha \in dQ^i$).

4 Public announcement updates and questions

Let us put things together. We are interested in the behavior of questions in public announcement logic. If our background epistemic logic is multi-modal S5, then the formula $[Q^i]Q^i$ is valid, i.e.,

Fact 1. *Questions are successful formulas.*

In S5-models a question Q^i, which is askable in a state s, is askable in all states from the equivalence class sR_i. No 'cutting' of states in the model $\mathbf{M}$ forced by the public announcement of Q^i results in $(\mathbf{M}, s) \models Q^i$ and $(\mathbf{M}|_{Q^i}, s) \not\models Q^i$. Thus, a publicly announced question is commonly known (see Proposition 3). In other words there is no model and state such that $(\mathbf{M}, s) \not\models [Q^i]Q^i$.

Successful formulas have an important property: they do not bring anything new if they are announced repeatedly.

Fact 2. *Let ϕ be a successful formula. $[\phi][\phi]\psi \leftrightarrow [\phi]\psi$ is valid.*

The proof is straightforward: $[\phi][\phi]\psi$ is equivalent to $[\phi \wedge [\phi]\phi]\psi$ (Proposition 1), which is equivalent to $[\phi]\psi$, because of the validity of $[\phi]\phi$ (ϕ is successful).

It is no surprise that askable questions (in a state) are successful updates.

Fact 3. $(\mathbf{M}, s) \models Q^i$ iff $(\mathbf{M}, s) \models \langle Q^i \rangle Q^i$.

---

[6] $(\mathbf{M}, s) \models \neg?_i\{\phi, \neg\phi\}$ iff $(\mathbf{M}, s) \models A_i?\{\phi, \neg\phi\}$.

Logic of Questions

Whenever an agent publicly asks a question, it does not cause any change in her epistemic model, it remains askable until she gets some new information.

The last point we are going to talk about is the relationship between public announcement and answerhood. Whenever a question is (partially) answerable in a state, then there is a formula ϕ such that after a public announcement of ϕ the question becomes inaskable there. Let us return to our group of players. Ann has the Joker. Neither Bill nor Catherine know it. If Catherine publicly asks

"Who has got the Joker?",

Bill can infer: "I have not the Joker and Catherine does not know who has it, therefore Ann has it." Catherine's question was *informative* for Bill, it caused that the question "Who has got the Joker?", which was askable for Bill, became inaskable after Catherine asked it, even if her question was not (partially) answered. This leads us to the definition of *informative formula*.

Definition 5. A formula ϕ is *informative* (for an agent i) with respect to Q in $(\mathbf{M}, s)$ iff $(\mathbf{M}, s) \models Q^i \wedge \langle \phi \rangle \neg Q^i$.

Contrary to partial answerhood (see the commentary below Proposition 4) there need not be any logical connection between an informative formula and direct answers to the question. The informativeness can be forced by the shape of the particular model.

In a multiagent epistemic approach we can understand a question as a 'task' (or a 'problem') to be solved by a particular group of agents. Communication is one of the basic tools of a group searching for a solution to a problem (an answer to a question) and asking questions is one of the essential parts of this communication. Questions may not only be about a fact *Who has got the Joker?*, but also about knowledge *Do you know who has got the Joker, Catherine?* or about a question. For example, Bill has a problem (he does not know who has the Joker) and wants to know if Catherine has the same problem. So he can ask *Would you ask 'Who has got the Joker?', Catherine?*. In fact Bill asks Catherine whether the question *Who has got the Joker?* is a reasonable (askable) question for her:

$$?_b\{?_c\{J_b, J_a\}, \neg ?_c\{J_b, J_a\}\}.$$

It is a yes-no question. The first direct answer means that Catherine would ask *Who has got the Joker?*, i.e., the question $?_c\{J_b, J_a\}$ is askable for her. The second direct answer means, this question is not askable for her, which according to Proposition 4 means that Catherine can (at least partially) answer that question.

5 Conclusion

We have presented the first step in the application of questions in the framework of dynamic epistemic logic. There are still many things to be done. In particular, we worked with the 'classical' public announcement logic based on the modal logic S5. We would like to use weaker systems of modal logic, so we will have to modify the definition of public announcement.

We would also like to study more deeply the relationship between group knowledge modalities and the problem of a search for an answer in a group of agents. We defined the askability and answerhood conditions for a single agent, but we can generalize this for a group of agents: it seems to be reasonable to require that a question is (partially) answered (for a group) if a (partial) answer is common knowledge (in this group). The dynamic aspect of common knowledge is mirrored in *relativized common knowledge*, see (Ditmarsch et al., 2008, section 7.8). Another interesting problem is under which conditions a group is able to find an answer to a question (without additional external information). This problem is closely connected to the notion of *implicit* (or *distributed*) knowledge.[7]

Michal Peliš and Ondrej Majer
Institute of Philosophy, Academy of Sciences of the Czech Republic
Jilská 1, 110 00 Praha 1
pelis@ff.cuni.cz, majer@site.cas.cz
http://logika.flu.cas.cz

References

Belnap, N., & Steel, T. (1976). *The Logic of Questions and Answers*. Yale.

---

[7]See, e.g., (Fagin, Halpern, Moses, & Vardi, 2003) for the corresponding definitions.

Benthem, J. van, & Minică, Ş. (2009). Toward a dynamic logic of questions. In X. He, J. Horty, & E. Pacuit (Eds.), *Logic, Rationality, and Interaction* (pp. 27–41). Springer.

Ditmarsch, H. van, Hoek, W. van der, & Kooi, B. (2008). *Dynamic Epistemic Logic*. Springer.

Fagin, R., Halpern, J., Moses, Y., & Vardi, M. (2003). *Reasoning about Knowledge*. MIT Press.

Genot, E. J. (2009). The game of inquiry: the interrogative approach to inquiry and belief revision theory. *Synthese*, *171*, 271–289.

Groenendijk, J., & Stokhof, M. (1990). *Partitioning logical space* (Annotated handout). ILLC, Department of Philosophy, Universiteit van Amsterdam. (Second European Summerschool on Logic, Language and Information; Leuven, August 1990; version 1.02)

Groenendijk, J., & Stokhof, M. (1997). Questions. In J. van Benthem & A. ter Meulen (Eds.), *Handbook of Logic and Language* (pp. 1055–1125). Amsterdam: Elsevier.

Harrah, D. (2002). The logic of questions. In D. Gabbay & F. Guenthner (Eds.), *Handbook of Philosophical Logic* (Vol. 8, pp. 1–60). Kluwer.

Hintikka, J., Halonen, I., & Mutanen, A. (2002). Interrogative logic as a general theory of reasoning. In D. Gabbay, R. Johnson, H. Ohlbach, & J. Woods (Eds.), *Handbook of the Logic of Argument and Inference* (pp. 295–337). Elsevier.

Peliš, M. (2008). Consequence relations in inferential erotetic logic. In M. Bílková (Ed.), *Consequence, Inference, Structure* (pp. 53–88). Faculty of Arts, Charles University.

Peliš, M. (2009). *Epistemic logic with questions*. (Available from http://web.ff.cuni.cz/~pelis/)

Tichý, P. (1978). Questions, answers, and logic. *American Philosophical Quarterly*, *15*, 275–284.

Wiśniewski, A. (1995). *The Posing of Questions: Logical Foundations of Erotetic Inferences*. Kluwer.

Wiśniewski, A. (2001). Questions and inferences. *Logique & Analyse*, *173–175*, 5–43.

Tales of Explosions
Andreas Pietz*

1 Introduction

In the discussion with David Lewis and others about how we reason in the face of inconsistency, Graham Priest reverted to a very interesting strategy: He wrote a story, "Sylvan's Box", about an inconsistent object, a box that is both empty and not empty. He then argued that the logic that the reader has to adopt to grasp what is going on in the story cannot be classical, because that would lead the reader to infer things that are clearly not true in the story. The problematic rule of inference is known as Ex Contradictione Quodlibet, or, more graphically, Explosion. It allows one to infer anything one wants from a contradiction. The invalidity of this rule is the defining feature of a paraconsistent logic, and so, in arguing that the reader of "Sylvan's Box" does not infer anything what so ever, Priest shows that the reader is reasoning paraconsistently. I take this to be a highly successful strategy to motivate paraconsistency, and I will argue that it might even be more successful than Priest himself thinks. The claim I wish to make is that the classical logician, were he to try to write a story that does for classical logic what Priest's story does for paraconsistent logic, would run into a phenomenon that philosophers of literature have called "imaginative resistance".

2 Sylvan's Box

2.1 The story

Priest's largely autobiographical story goes as follows: After the sudden death of his long time friend and colleague, Richard Sylvan (Routley), Priest drives to Sylvan's remote home to start to work through

---

*The research for this paper was partly funded by grant HUM2008-FFI04263 awarded by the Spanish Ministerio de Ciencia e Innovación. I'd like to thank Roberto Ciuni, Genoveva Marti, Chris Mortensen, Graham Priest, Sven Rosenkranz, Heinrich Wansing, the members of the LOGOS group in Barcelona and the audiences at Logica in Hejnice, GAP 7 in Bremen, Enfa 4 in Evora and two talks I gave in Dresden and Barcelona for very helpful discussion.

his *Nachlass*. In between piles of papers he finds a little box with the words "Inconsistent Object" written on it. As he inspects it he finds it, incredibly enough, both empty and not empty. Inside there is a little figurine and, at the same time, there is no figurine at all. This is not an optical illusion, the box is an object with perfectly contradictory properties. Priest and Nick Griffin, who is also present, discuss what to do with such a remarkable find, and they find themselves in two minds: they don't want to miss this chance of proving to the world that true contradictions exist, but neither do they want the box to fall into the wrong hands. The story ends on a truly inconsistent note: Priest takes the box with him when he leaves Sylvan's house, while Griffin buries it in Sylvan's garden.

After this inconsistent ending, the reader is presented with a questionnaire about the story that tests whether the reader has understood what is going on in the story. Priest then goes on to draw ten morals from this story about inconsistency, interpretation and the nature of impossible worlds and impossible objects.

What is especially important for this piece is the following: The reader of Sylvan's Box can understand and reason about the story, even though it is inconsistent.[1] Furthermore, he is lead to reason in a certain way, i.e. paraconsistently.[2] This can be seen by looking at his answers to the questionnaire; he would have been wrong to reason in a different way. For example, if he concluded via Explosion that the box was shot off to the moon in the end of the story, he would not have read the story correctly.

However, neither would he have understood what was going on by following Lewis's suggestion how to deal with inconsistent bodies of information in (Lewis, 1978). Here the idea would be to break up the story into (maybe maximally) consistent chunks and only draw inferences from those. But that way the reader would not only miss the whole point of the story, he would be drawing false inferences. Asked why Priest was so excited in the story, he would have to either

---

[1] C. Mortensen even demonstrates that he can visualize what is going on. See http://www.hss.adelaide.edu.au/philosophy/inconsistent-images/sylvans/.

[2] Paraconsistency, i.e. the invalidity of Explosion, is not to be confused with dialetheism, a metaphysical doctrine also defended by Priest. According to dialetheism, there are true contradictions. Of course, the story is as little proof of dialetheism as Doyle's stories prove the existence of Holmes, but proving dialetheism is not at all the aim of the story.

answer it was because he found the box to be empty, or alternatively answer that it was because the box wasn't empty. But neither is right, Priest was excited because the box was both empty and not empty.

2.2 Lewis's reaction to the box

In any case, Lewis had in mind everyday cases of mistaken beliefs leading to inconsistent belief sets, and he never meant to suggest that strategy in the case of obviously contradictory stories. His position in "Truth in Fiction" was that

> "anything whatever is vacuously true in an impossible fiction. That seems entirely satisfactory if the impossibility is blatant: if we are dealing with a fantasy about the troubles of the man who squared the circle, or with the worst sort of incoherent time-travel story. We should not expect to have a non-trivial concept of truth in blatantly impossible fiction [...]" (Lewis, 1978, p. 275)

Sylvan's Box, however, had the effect of forcing Lewis to give in at least somewhat, as is evidenced by this quote from a letter he wrote to Priest (reprinted in (Priest, Beall, & Armour-Garb, 2004, p. 176)):

> "I'm increasingly convinced that I can and do reason about impossible situations. ("Sylvan's Box" played a big part in persuading me.)"

He does not go so far as admitting that the logic one has to employ in reasoning about stories such as "Sylvan's Box" has to be paraconsistent, though. He picks up the distinction between "blatant" and "subtle" impossibilities he had made earlier and argues that the issue of paraconsistency only arises in connection with blatant impossibilities. Even though he himself admits to be very unclear about the distinction, he must have been taking the impossibility of "Sylvan's Box" to be a subtle one, because he calls the mention of paraconsistency here "off the topic". Had it not been for his untimely death, he might have worked his view out more fully, but as it stands I have to say I can't see in what way the inconsistency of "Sylvan's Box" is not blatant, and neither can I fathom how one might appreciate the intuitive thrust of the story and not take it to be about paraconsistency.

3 Story-Inducable Logics

One might try to make all this into a very strong argument for paraconsistent logic. At the very least, we have to be able to employ a paraconsistent logic to deal with this story, and we don't seem to have much of a problem with that. If the logic we are employing is not paraconsistent from the outset, then it seems strange that we can make the transition so easily. Priest actually doesn't take that line of argument. He suggests that for any logic whatsoever, a story could be told that would make it seem the only suitable logic for this story. That is what he did with *Sylvan's Box* for paraconsistent logic, and he suggests one could write a story about a man having entered a room through two doors without having entered through one in particular, a story that is supposed to get the reader to reason along the lines of quantum logic.

It is interesting to follow this hypothesis through and and investigate what kinds of logic can be forced onto the reader in a similar way. I will write that a story induces a given logic when this kind of maneuver succeeds. What would a story look like that induces fuzzy logic, or one that *induces* intuitionistic logic? I think these are fascinating questions, but in this paper I'd like to experiment with a seemingly easier case, namely classical logic. Is there a story to be told to a paraconsistentist that would get him to reason classically about what is going on in the story?

4 Inducing Classical Logic

4.1 The indirect Strategy

The interesting difference between classical logic and a paraconsistent one is that classical logic validates Explosion, the principle that from a contradiction any proposition whatsoever follows. So let's see how one would go about telling a story that induces a logic that validated Explosion. One way would be a consistent story that exhibits typical inferential moves that paraconsistent logicians are traditionally having trouble with, such as Disjunctive Syllogism and Modus Ponens. In a second step the defender of Explosion would challenge the paraconsistentist to come up with a workable logic that validated all the inferences drawn in the story and didn't validate Explosion.

Even though Priest hasn't made any suggestion about which logic in fact is employed by the readers of his story, we may speculate that he would recommend his LPm. LPm is introduced in the second edition of his *In Contradiction* (Priest, 2006a), and is an improved version of his Logic of Paradox (LP), the logic that might well be the most famous in the paraconsistent family. While LP might in fact be outmaneuvered by a suitably constructed story, LPm can deal with those challenges. That is because LPm is a so-called *adaptive logic*, which means that it behaves classically in a consistent environment and switches to the paraconsistent logic LP only when it comes across a contradiction. Therefore, it thwarts all attempts to challenge the paraconsistentist with a consistent story, as there will be no difference in the inferences validated by LPm and classical logic in such a story.

4.2 The direct strategy

More satisfactory, in any case, for the classical logician would be to come up with a story that induces Explosion directly. What we want is a story that makes the reader infer everything. To test this we might give him a questionnaire much like Priest did. Some care has to be taken with the phrasing of the questions, however. If a question is, e.g., "Can pigs fly in the story", the answer we would hope for would be "Yes". But even if we succeeded in making the reader infer everything, he might still answer "No", meaning that in the story pigs cannot fly, which of course should also be true. Therefore, let's only ask questions of the form "Is it true in the story that X". Whatever X may be (and we should include a wide variety of propositions in the questionnaire), we should hope for the answer "Yes" and nothing else.

But even if we get the reader to check the "Yes" box after every question. we cannot be sure to have succeeded in inducing Explosion. We have to make sure that the only way to conclude that every answer to the questions is "Yes" is indeed by way of the principle of Explosion. Take the minimal (or rather maximal) "story":

> Everything is true.

Of course, the answers to our questions about this story will come out "Yes", but Explosion is neither the only nor the most straightforward way to infer these answers. We have to come up with something more refined.

4.3 Three Attempts

One thing is clear: The story has to sport an inconsistency, and probably a quite blatant one at that. We already have such a story to work with, namely Sylvan's Box. Let's try to turn it into a story inducing Explosion.

As Explosion tells us that anything follows from a contradiction, one might simply try to append all kinds of random statements to the story. E.g.:

Alternative Ending 1:

> "[Sylvan's Box] As I [i.e. Priest, the story is told from the first person perspective] drove down into town with the box in the trunk of my car, I came about a flock of pigs flying around my car, and the moon was made of green cheese. To my further surprise I realized that the sum of five and seven was fourteen. And eleven as well."

Append enough of these, and a logician might get what is meant to be going on, but surely this isn't much more than an insider joke. It's another interesting project to construct stories such that a trained logician could find out what logic is meant to be characterized. But such a logical detective story is not what we are after.

We want a story that makes logicians and laymen alike infer classically. No layman reading the story would take those statements to be conclusions of the former contradiction. The reason is simply that one cannot make statements be taken as consequences of each other simply by writing them in sequence.

Then what does it take to turn a series of statements in a story into a logical consequence from premises to conclusions? Well, at this point I have to say that I can't think of any other way to pull this off than to state the aim outrightly, for example thus:

Alternative Ending 2:

> "[Sylvan's Box] As I drove down into town, I checked the contents of the box again. Sure enough, it was still empty. And sure enough the box still contained the little figurine.

Tales of Explosions

> Therefore, a flock of pigs turned up on the horizon and flew towards me, soon to be circling around my car."

I suggest that this strategy will not be successful. To my mind, the problems it runs into are the same that the literature knows under the name "imaginative resistance". In short, the reader, even if she has swallowed all that preceded the last sentence, is not likely to buy this last part. It's not that pigs can fly that she can't accept, but rather that this is supposed to be a logical consequence of the preceding contradiction. It's exactly the "therefore" that will upset her reading experience. It may be that she just isn't sure what the "therefore" is meant to express here, as one isn't accustomed to reading something like that in a short story. One may try to elaborate to get the point through:

Alternative Ending 3:

> "[Sylvan's Box] As I drove down into town, I checked the contents of the box again. Sure enough, it was still empty. And sure enough the box still contained the little figurine. This was clearly contradictory, and I knew what I had to expect to happen. It's a logical truth that everything follows from a contradiction.. And so it was with horror, but without real surprise that I saw a flying pig that was about to hit my windshield."

I submit that this isn't much better. I'm not sure that it is possible to tell a truly trivial story (one in which everything is true) that the reader can engage in, but in any case I think she will resist the claim that everything is to be inferred from the earlier contradiction. And any "normal" attempt to motivate Explosion further will fail, because it will need to invoke the principle that a contradiction is never true, which is clearly violated in any inconsistent story from the outset.

Of course, these attempted alternative endings won't make for the most engaging literary experience. After all, much of the persuasive strength of Sylvan's Box lies in it's being a well written short story. However, no matter how well crafted one's story might be, I don't see how the reader can be made to draw explosive inferences without the narrator telling him to, and I think in every such case a feeling of resistance will be felt.

5 Imaginative Resistance

5.1 The puzzles about morality and humor

The term "imaginative resistance" came up in the discussion about the problems that arise in relating deviant morality or humor in stories. A story might have a mob of outraged commuters killing a couple that is clogging up traffic on the freeway. (This example is the story that Brian Weatherson tells to illustrate the phenomenon in (Weatherson, 2004)). It's no problem at all to tell this in a way that engages the reader enough to have him "buying" it, ie to get him to hold the given events as true in the story. However, it is very hard indeed to do the same with a moral judgment about the situation. Just because the narrator tells you that the mob was right to kill the couple doesn't make it true, not even true in the story. The reader's imagination will revolt at that point, hence the name "imaginative resistance". In Weatherson's words, "we refuse, fairly systematically, to play along with the author here". (Weatherson, 2004, p. 2)

The same is said to hold in he case of humor. There is no problem in having a character tell a joke that is not funny, but it's hard to then make it part of the story that it actually *is* funny. It is far from clear what the reason for these refusals of cooperation on the part of the reader is, and none of the attempts to explain the phenomenon in the literature is wholly convincing. Personally, I take the accounts that focus on the role of the narrator to be the most promising. In cases of imaginative resistance, the narrator suddenly becomes a salient agent, and moreover one who has overstepped the boundaries of his role. The narrator is supposed to relate the factual events of a given story, and is not supposed to give moral judgements or decide wether a joke is funny or not. A good illustration likens the narrator of a story to a foreign correspondent reporting from a distant country. As long as that person sticks to the facts, he is in a much better epistemic position than we are, and we tend to believe him as long as we have no independent reason to doubt his words. As soon as he starts to dish out moral judgements, though, he is not in a better position than we, and we react with a sense of "Well, that's what *you* think, but I beg to differ!". We pride ourselves to be able to make our moral judgement on our own, and we don't like to be told what to think in these matters, neither by a foreign corespondent nor by the narrator of a fictional story.

What I'd like to suggest is that the weirdness of the examples I gave above is of the same kind as the cases where deviant moral judgements or divergent senses of humor are attempted to be pushed onto the reader. It would be interesting to see which of the explanations that have been offered in the moral and jocular cases might carry over to the logical case, but my point is sufficiently made by flagging the phenomenon. I think that the reader of one of my proposed stories will respond in exactly the same way as the reader of Weatherson's story about the avenging commuter. He might accept all that happens factually, but when it comes to judging what follows from what, he will insist on doing the inferring for himself and reject any inferential moves the narrator makes that he doesn't accept to be valid.

Now, one last obvious observation: It is of course only when the moral judgements of the narrator *clash* with our own that we resist them. If Weatherson's narrator had gone on to tell us that it was wrong of the man to kill two people just because they were causing a traffic jam, it might have struck us as a strange bit of storytelling. However, we wouldn't have resisted the judgement and concluded that the murder was indeed justified, just because we don't like to be told what to think. That would be a bit juvenile, after all.

If the case of logical resistance indeed is of the same sort as the other kinds, then it would be hard for anyone to argue that a logic that we feel imaginative resistance toward is actually the one that we employ outside of story-reading. It would be the only case in which something like this is happening.

6 Concluding Remarks

I think that the field of research that Sylvan's Box opened up is a very exciting one. It would be great to see what kinds of logics can be induced by appropriately designed stories. On the other hand, it would be interesting whether the special kind of resistance that some logical stories face can shed light on the moral cases. In any case I take it to be a further challenge for the classical logician to explain why not only we are perfectly able to reason paraconsistently, but why we moreover don't like to be told to reason classically.

Andreas Pietz
Universitat de Barcelona
LOGOS Research Group
andreas.pietz@gmail.com

References

Hanley, R. (2004). As good as it gets: Lewis on truth in fiction. In F. Jackson & G. Priest (Eds.), *Lewisian themes*. Oxford: Oxford University Press.

Lewis, D. (1978). Truth in fiction. *American Philosophical Quarterly*, *15*(1), 37–46.

Lewis, D. (1982). Logic for equivocators. *Noûs*, *16*(3), 431–441.

Priest, G. (1999). Sylvan's box: A short story and ten morals. *Notre Dame Journal of Formal Logic*, *38*(4), 573–582.

Priest, G. (2006a). *In contradiction* (2nd ed.). Oxford: Oxford University Press.

Priest, G. (2006b). *Towards non- being: the logic and metaphysics of intentionality*. Oxford: Oxford University Press.

Priest, G., Beall, J., & Armour-Garb, B. (Eds.). (2004). *The law of non-contradiction: New philosophical essays*. Oxford: Oxford University Press.

Weatherson, B. (2004). Morality, fiction, and possibility. *Philosopher's Imprint*, *4*(3), 1–27.

On Attempting

Tomasz Placek*

1 Problem

Alicja (3 yrs) attempted to break her mug at our breakfast yesterday. How did she go about this? She slowly moved the mug towards the edge of our table until it tilted and began to fall down. It did not break, however, as I caught it before it hit the floor. I scolded Alicja for *attempting* to break her mug, but she did not agree. So we had a row. To convince her, I gave her a short talk on agency and in particular, on the *stit* theory, but I cannot call it a success. I nevertheless report on our exchange as (I think) it casts light on two questions: (i) to what extent is attempting an intentional concept? And (ii) how far can one get in analyzing it in the stit framework?

In what follows I make three assumptions. First, "to attempt" is a so-called accomplishment verb, which means that its meaning changes with tenses. A sentence in the past tense, like "I attempted to prove this lemma yesterday", implicates that I failed to prove it. My assertion, "I am attempting to prove this lemma", however, is consistent with my success as well as with my failure. In this text I leave aside this complexity and arbitrarily assume the past-tense meaning of "to attempt", i.e., I take it that attempting is always unsuccessful. Second, I assume an idealization that attempting occurs at point-like moments. Third, although English grammar prescribes the form "α attempts to break her mug", I suggest a little linguistic reform here: I will write "α attempts that her mug will be (is) broken". In other words, I assume that "α attempts that ... " is a propositional operator, my aim being to find adequate truth-conditions for it.

2 *Stit*, branching time (BT), Agents, and Choices

Stit is an acronym for the phrase "see to it that" which, as Belnap, Perloff and Xu (2001) have suggested, might serve to draw a distinc-

---

*This work is supported by MNiSW research grant 3165/32. It extends my earlier paper, (Placek, 2009, in Polish), and corrects its errors. For many insightful comments on attempting and the stit theory, I am very grateful to Nuel Belnap.

tion between agentive and non-agentive sentences. To give an example, the sentence "I am going to Prague" is ambiguous, as in some contexts it is agentive and in others it is not. But, if its assertion in a given context can be paraphrased by using the stit form — that is as "I see to it that I am going to Prague" — the sentence is agentive; otherwise it is not.

The stit theory presents the task of understanding agency by explaining the phrase "see to it that". The phrase is assumed to be represented by a propositional operator, written as *stit*; the task amounts to providing a semantics for this operator. The explanation does not appeal to mental concepts, like the agents being conscious, or driven by some intentions, or constrained by moral norms and values. This does not mean that an agent's actions do not have such aspects. The idea is that basic features of actions can be understood without taking recourse to mental concepts, and if there is a need for such a concept, the stit theory will lend itself to an appropriate supplementation.

The semantics for stit is based on the theory of *branching time* (BT), which was put forward by Prior (1967). A language $\mathcal{L}$ considered has, besides the *stit* operator, the tense operators P (it was the case that) and F (it will be the case that), the operators of historical possibility and historical necessity, respectively, poss and sett, the tense indexicals like 'now' or 'tomorrow', and the truth-functional connectives $\vee, \wedge, \neg, \rightarrow$ of disjunction, conjunction, negation, and (material) implication, respectively.

It is assumed that the atomic sentences of our language $\mathcal{L}$ are in the present tense and significantly tensed. We read "F: Kasia reads a book" as "Kasia will read a book" and analogously for the past tense operator P. The sentence "sett: P: Kasia goes to Warsaw" is read as "It is (already) settled that Kasia has gone to Warsaw". In a similar vein, "poss: F: Kasia goes to Warsaw" is rendered as "It is (still) possible that Kasia will go to Warsaw".

Turning to the branching time theory, its model $\mathcal{W} = \langle W, \leqslant \rangle$ is a non-empty and partially ordered set, subject to the condition of no backward branching —

$$\forall x, y, z \in W (x \leqslant z \wedge y \leqslant z \rightarrow x \leqslant y \vee y \leqslant x),$$

and the condition of historical connection, i.e.,

$$\forall x, y \in W \exists z \in W : (z \leqslant x \wedge z \leqslant y).$$

Since histories are defined as maximal chains in W, the last condition implies that every two histories intersect, which justifies the condition's name. Elements of W are called "events", which is rather unintuitive since a BT event is an instantaneous slice of the universe, i.e., a class of point events simultaneous with a given point event. (As the theory assumes absolute simultaneity, it is pre-relativistic.) Finally, $x \leqslant y$ means that y belongs to a possible future of x, or (equivalently) that x is in the past of y. We write $Hist$ for the set of histories in $\mathcal{W}$.

There is a family of undividedness relations $\equiv_e$ on Hist, parametrized by events $e \in W$ and defined as $h_1 \equiv_e h_2$ iff $\exists e' e' \in h_1 \cap h_2 \wedge e < e'$. It is straightforward to see that $\equiv_e$ is an equivalence relation on $H_{(e)} := \{h \in \text{Hist} \mid e \in h\}$. Thus, $\equiv_e$ induces the partition Π_e of $H_{(e)}$.

Some BT models allow for the introduction of a set Ins of instants, thought of as instantaneous moments at which events occur. It is required that (i) every element of Ins and every history intersect at a single event, and that (ii) Ins respects ordering $\leqslant$ on W, which amounts to the following conditions. For every $e_1, e_2, e'_1, e'_2 \in W$, every $i_1, i_2 \in \text{Ins}$, every $h_1, h_2 \in \text{Ins}$:

$$h_1 \cap i_1 = h_1 \cap i_2 \to h_2 \cap i_1 = h_2 \cap i_2 \tag{1}$$

$$(e_1 \in h_1 \cap i_1 \wedge e_2 \in h_1 \cap i_2 \wedge e_1 < e_2 \wedge e'_1 \in h_2 \cap i_1 \wedge \\ \wedge e'_2 \in h_2 \cap i_2) \to e'_1 < e'_2. \tag{2}$$

We write $i(e)$ for the instant determined by event e.

Turning to the BT semantics, Prior's main semantical idea is to take for evaluation points the event-history pairs in which the event belongs to the history. Such a pair is written as e/h. A BT semantical model $\mathfrak{M} = \langle \mathcal{W}, I \rangle$ for our language $\mathcal{L}$ consists of a BT model $\mathcal{W}$ and an interpretation function $I \colon \text{Atoms} \to \mathcal{P}(W)$, where Atoms is the set of atomic formulas of $\mathcal{L}$. The evaluation is defined as follows:

$$\text{For } A \text{ an atomic formula: } \mathfrak{M}, e/h \models A \text{ iff } e \in I(A); \tag{3}$$

$$\mathfrak{M}, e/h \models \neg \varphi \text{ iff it is not the case that } \mathfrak{M}, e/h \models \varphi; \tag{4}$$

$$\text{and similarly for the other classical connectives;} \tag{5}$$

$$\mathfrak{M}, e/h \models F \colon \varphi \text{ iff there is } e' > e \text{ such that } \mathfrak{M}, e'/h \models \varphi; \tag{6}$$

$$\text{and analogously for a formula with } P \text{ as the main operator;} \tag{7}$$

$$\mathfrak{M}, e/h \models At_i \varphi \text{ iff } \mathfrak{M}, e'/h \models \varphi \text{ where } e' \in i \cap h; \tag{8}$$

$$\mathfrak{M}, e/h \models \text{sett} \colon \varphi \text{ iff for every } h' \text{ if } e \in h' \text{ then } \mathfrak{M}, e/h' \models \varphi. \tag{9}$$

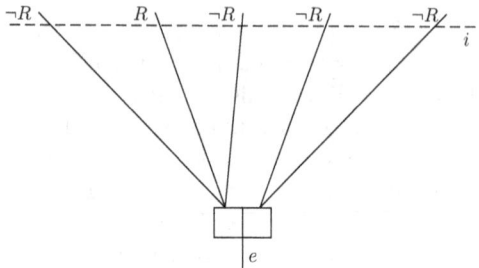

Figure 1: At e/h agent α risks that $At_i R$.

poss: φ is defined as equivalent to ¬: sett: ¬φ.

To represent agents' actions, one first postulates a non-empty set Agents. Then choices available to Agents at events from W are coded by a "choice function" $Choice$: Agents $\times W \longrightarrow \mathcal{P}(\text{Hist})$ subject to the conditions: (i) $Choice(\alpha, e)$ is a partition of $H_{(e)}$ and (ii) if $h \in Choice(\alpha, e)$ and $h \equiv_e h'$, then $h' \in Choice(\alpha, e)$. We will write $Choice_e^\alpha$ for $Choice(\alpha, e)$ and $Choice_e^\alpha \langle h \rangle$ for the element of $Choice_e^\alpha$ determined by h (this is defined provided that $h \in H_{(e)}$).

3 First idea: attempting is risking

Our analysis of attempting will proceed in stages. The first idea is to link attempting to risking. Namely, α attempted that the mug be broken (R) at instant i means that she risked that R at instant i. We suggest analyzing this in the following manner: there was an option available to α in which certainly $\neg R$ (the mug is not broken) at i, but α chose another option which allowed that the mug be broken at i (see Figure 1). In symbols our proposal reads:

$\mathfrak{M}, e/h \models \alpha$ risks that $At_i R$ iff

1. $\exists h' \in \text{Hist}: h' \in Choice_e^\alpha \langle h \rangle \wedge \mathfrak{M}, e/h' \models At_i R$,

2. $\exists A \in Choice_e^\alpha \; \forall h'' \in A: \mathfrak{M}, e/h'' \models At_i \neg R$.

Risking so described relies on forsaking a sure thing in favor of a not-so-sure thing. The concept is too strong, however. By the above explication, I should not say that Alicja risked breaking her mug (even though she put it on the very edge of our table), if I believe that, if left alone, the mug could still break, either for some physical reasons, or because of the workings of other agents — our cat, for instance. A question thus arises with respect to what do we risk? In particular, should we require of a state that a risky agent forsake that it has a sure-thing characteristic? We will return to this question in the next section.

Note also that in this analysis, attempting is ubiquitous. By deciding to spend this day in bed, I would ensure that quite a few things would (not) happen. For instance, I would ensure that my milk pot would now be intact. Yet, I got up, I decided to boil milk for Alicja, put the pot on our stove, and... forgot about it. The pot would have burned if my wife had not intervened. So, by deciding to get up, I attempted to burn the pot, and most likely I attempted an indefinite number of other things.

It seems we may improve on our first analysis by combining it with some concept of guaranteeing: I attempted that R because first I risked that R and second, although later I could still guarantee that $\neg R$, I did not do it. So we turn next to guaranteeing.

4 Guaranteeing I

Consider the four diagrams of Figure 2. In each diagram at e/h it is true that agent α risks that R at instant i. In the top two diagrams, after taking this risk (by picking the "left" option), α can ensure at a later event e' that $\neg R$ at i. One might got the impression, however, that guaranteeing is achieved by an accidental fact: α's choices at e' (as coded by $Choice^{\alpha}_{e'}$) combine nicely with possibilities on which α has no influence. In the top-right diagram, α has no way of choosing the third history from the left rather than the first or the second, but fortunately, in this history we have $\neg R$ at i. And in the top left diagram, there is simply no history which α cannot choose.

The bottom two diagrams, however, contain histories that destroy guaranteeing, as they have R at i, and α cannot exclude them. Before addressing this problem, we need to decide what should be on the right-hand side of those diagrams that illustrate guaranteeing. This amounts to deciding with respect to what we risk or attempt. In other words, if Alicja attempted to break her mug by taking such and such an action, what should we say would happen to the mug if she did not take this action? What should be demanded of choice $A \in \mathit{Choice}_e^\alpha$, which α does not take at e/h, that is, $A \neq \mathit{Choice}_e^\alpha \langle h \rangle$?

(i) Surely the mug would not break. Formally,

$$\forall h' \in A \forall e'(e' \in i \cap h' \to e'/h' \models \neg R.$$

Clearly, this condition is too strong. We may still believe that α risks breaking the mug, even though we think that without α's action the mug's integrity is nonetheless endangered by our cat, a draught, or Mary's action. A possible obstructive action of some other agent does not get α off the hook, so to speak.

(ii) Other agents, working separately or jointly, could ensure that the mug is not broken. The idea is that option A contains a history in which the mug is broken at i, but other agents have choices available that would lead to $\neg R$ at i in the appropriate history. This makes α's attempting dependent on choices available to other agents, and this is not a good idea.

(iii) α herself might guarantee that $\neg R$. Despite having been able at e/h to guarantee that $\neg R$ at i, she chose at e another option such that she could not later guarantee that $\neg R$ at i. However, in the present reading we allow for the obstructive actions of other agents and obstructive indeterministic workings on the part of nature even in the "guaranteeing option". The idea here is to understand guaranteeing in a rather relaxed, or counterfactual way: if there were no obstructive workings by other agents or nature, α could guarantee $\neg R$ at i. Thus the task is to put in brackets the obstructive workings of other agents and nature. To analyze this idea, we will employ the concept of simple strategies of Belnap, Xu, and Perloff (2001), which we even further simplify for the present purposes.

5 Guaranteeing II and simple strategies

Our aim now is to analyze, in the framework of simple strategies, the phrase "α can guarantee that Q at instant i", where we assume the relaxed interpretation of "guaranteeing" as indicated just above.

On Attempting

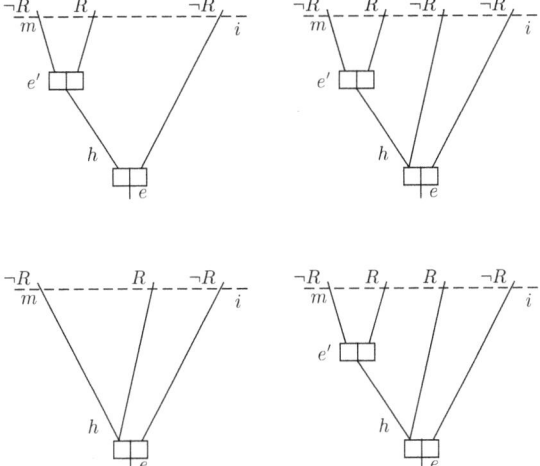

Figure 2: Guaranteeing I.

A strategy s available for $\alpha \in$ Agents is a partial function on W such that:[1]

$$\text{if } e \in \text{Dom}(s) \text{ then } s(e) \in \text{Choice}_e^\alpha.$$

We say that s is a strategy for α in field M if $\text{Dom}(s) \subseteq M \subset W$. To define simple strategies, one needs the following notions:

1. Admissibility: s admits e iff

$$\forall e_0 (e_0 \in \text{Dom}(s) \wedge e_0 < e \rightarrow e \in \bigcup s(e_0));$$

2. Primariness: strategy s is primary iff $e \in \text{Dom}(s) \rightarrow s$ admits e;

3. Backward closure: for strategy s for α in field M, we say that s is backward closed in M iff

$$\forall e_1, e_2 \ (e_2 \in \text{Dom}(s) \wedge e_1 < e_2 \wedge e_1 \in M \rightarrow e_1 \in \text{Dom}(s));$$

4. Simplicity: a strategy s for α in M is simple iff it is primary and backward closed in M.

---

[1] To simplify, we assume here that a strategy is strict in the sense that $s(e) \in \text{Choice}_e^\alpha$, rather than $s(e) \in \mathcal{P}(\text{Choice}_e^\alpha)$.

Consider now the set of histories such that at the pair "event e, history", the sentence "At_iQ" is true, i.e.,:

$$H_e^{At_iQ} = \{h \in H_{(e)} | e/h \models At_iQ\}.$$

With this set, we explain the phrase "α can guarantee that Q at instant i" as follows:

$e/h \models \alpha$ can guarantee that Q at instant i iff there is a simple strategy s for α in $M = \bigcup Choice_e^\alpha \langle h \rangle$ such that: $\bigcap_{x \in \text{Dom}(s)} s(x) \subseteq H_e^{At_iQ}$.

Let us return to Alicja's actions at the table. The situation is depicted by Figure 3, with white boxes representing Alicja's choices and a shadowed box indicating the choices of some other agent, say, our cat. Can Alicja guarantee that the mug is not broken, $\neg R$, at instant i? At e/h it is not true that she can guarantee that $\neg R$ at i: there is no strategy for Alicja with field $M = \bigcup Choice_e^{Alicja} \langle h \rangle$ such that $\bigcap_{x \in \text{Dom}(s)} s(x) \subseteq H_e^{At_i \neg R}$. In other words, if Alicja selects the "left" option at e, she cannot guarantee that the mug is not broken. However, she can guarantee this, if she chooses the "right" option at e, because there is a strategy for Alicja with field $M = \bigcup Choice_e^{Alicja} \langle h'' \rangle$ such that $\bigcap_{x \in \text{Dom}(s)} s(x) \subseteq H_e^{At_i \neg R}$ for arbitrary $h'' \in Choice_e^{Alicja} \langle h' \rangle$. The strategy relies on taking the "right" option at e and then the "left" option at e''. Note that in the "right" option the verdict is that Alicja can guarantee ..., despite a possible obstruction by our cat at e'.

6 A final analysis

We have come to our final analysis:

$e/h \models \alpha$ attempts that At_iR iff

1. $e/h \not\models At_iR$, and

2. $e/h \not\models \alpha$ can guarantee that $\neg At_iR$, and

3. there is $h' \in H_{(e)}$ such that: $e/h' \models \alpha$ can guarantee that $\neg At_iR$.

My story about Alicja's behavior at the table now goes like this: At e/h Alicja attempted to break her mug in 1 sec. How did she do that? At e/h she placed her mug on an edge of our table. At this event it was not yet settled whether the mug would break: both

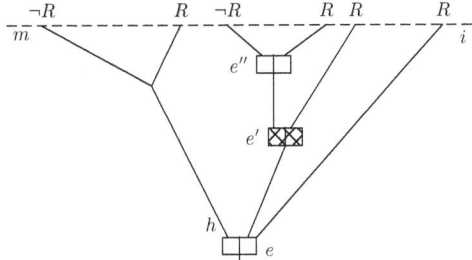

Figure 3: Alicja and our cat at the table. White and shadowed boxes represent Alicja's actions and our cat's actions, respectively.

scenarios were possible. Alicja had other options at e: she might have left her mug in front of her, where it stood. Yet, even if she had chosen to leave the mug, it might still have broken, which might have happened spontaneously (the farmost right history), or not. In the latter scenario, our cat will come to the mug (event e). The cat will face a choice, to break the mug outright or to play with it. In the latter option, Alicja has a choice to save the mug or not (at e''), by chasing the cat away or not.

7 Objections

Our analysis leaves aside the intentional aspects of agents' actions. The two objections presented below try to show that, as a result, the analysis is both too wide and too narrow.

Consider a careful pedestrian on a sidewalk, choosing whether or not to start crossing the street. As there is no incoming traffic to be seen at this moment, he elects to briskly walk through the crossing. When he is in the middle of the road, however, with a cosmic velocity a mad car appears, heading directly towards him. Fortunately, the mad driver turns his vehicle left at the last moment, saving the pedestrian. Now, did the pedestrian attempt to get injured by deciding to cross the street? Clearly, given his option of not crossing the street, he could arguably guarantee that he be uninjured. He chose to cross the street, however, the consequence being that he could not guarantee that he be uninjured. So, by our analysis, the pedestrian attempted to get injured. But, intuitively speaking, he did not. "It was not his intention to get injured" comes as a natural comment.[2]

In the second story, a platoon of soldiers is surrounded by a dreadful enemy.[3] There is no way for them to escape the ambush. Nevertheless, the commanding officer gives the platoon an order to attack. As a result, all the soldiers die. The officer, later tried for foolhardily sending his soldiers to certain death, maintains a defense that by giving the order to attack, he *attempted* to get the platoon out of the ambush. So his fate depends on the truth at e/h of the sentence: "The officer attempts that the platoon be out of the ambush at some later instant i", where e is a moment of issuing the order and h is an arbitrary history in which the trial occurs. In symbols,

$$e/h \models \text{ the officer attempts that } At_i\, \text{Out}.$$

The prosecution, however, has solid evidence that it was impossible to escape the ambush. Also, on the basis of prior evidence concerning the enemy's cruelty, it is clear that if the soldiers had surrendered, the enemy would have killed them. Thus, well before issuing the order, there was no possible scenario in which the platoon survives. This means that in every history comprising the event of issuing the order the sentence $\neg$ Out is true at instant i. By the concept of guaranteeing, this means that at e/h the officer can guarantee that $\neg At_i$ Out. Accordingly, the second clause of our final analysis is not satisfied, so this analysis yields a negative verdict: the officer did *not* attempt to get his platoon out of the ambush.

Common sense might say the opposite, in particular, if the officer convincingly shows that (i) he thought (mistakenly) that the situation was not hopeless and (ii) he had a plan (unrealistic, as it was) of how to rescue the soldiers, and the order in question was a part of this plan. As far as I could learn, the court verdict might go both ways, depending on the nature of the officer's mistaken assessment of the situation.

My own views agree with the final analysis, however. The officer did not attempt to rescue the platoon. He thought (believed) that he was attempting to save the platoon, and, perhaps, one can even make

---

[2] I owe this example to John Kearns.
[3] This story comes from Tim Childers. I am indebted to Michał Araszkiewicz for explaining to me the definitions of attempting as they occur in the Polish criminal code.

a case that he was justified in believing that he was attempting to save the platoon. Nevertheless, my strong intuition is that, given the hopeless predicament, the officer was mistaken: he was not attempting to save the platoon.[4]

Dismissing the second story as a counter-example to the final analysis, I am left with the first story which shows that the analysis is too wide. The story, however, suggests clearly a condition to be added to our analysis: an agent's intention. The challenge (which we do not know how to address) is to express this condition in the stit framework, however. For the record, I nevertheless put down this idea as an informal, unofficial, and post-final analysis:

$e/h \models$ attempts that $At_i R$ iff

1. $e/h \not\models At_i R$, and
2. $e/h \not\models \alpha$ can guarantee that $\neg At_i R$, and
3. there is $h' \in H_{(e)}$ such that: $e/h' \models \alpha$ can guarantee that $\neg At_i R$, and
4. $e/h \models \alpha$ intends that $At_i R$.

Examples similar to our first story as well as criminal codes indicate that intending, planning, or even pre-meditating have some role in attempting. To obtain some clarity about what exactly this role is will be a task for a future project. Yet another task is to see if the recent stit logic for belief, desire, and intention by Semmling and Wansing (2009) can be harnessed to analyze the mental aspects of attempting.

Tomasz Placek
Department of Philosophy, Jagiellonian University
Grodzka 52, 31-044 Kraków, Poland.
tomasz.placek@uj.edu.pl

References

Belnap, N., Perloff, M., & Xu, M. (2001). *Facing the future*. Oxford: Oxford University Press.

---

[4]Of course, I am applying here a weak understanding of justification according to which we can have justified false beliefs.

Placek, T. (2009). O usiłowaniu. In J. Czetwertyński-Sytnik (Ed.), *Rozważania o filozofii prawdziwej* (pp. 105–115). Kraków: Wydawnictwo Uniwersytetu Jagiellońskiego.

Prior, A. (1967). *Time and modality*. Oxford: Oxford University Press.

Semmling, C., & Wansing, H. (2009). A sound and complete axiomatic system of bdi–stit logic. In M. Pelis (Ed.), *Logica Yearbook 2008* (pp. 193–210). London: College Publications.

"This sentence" is indexical.
The indexical variant of the Liar paradox and McTaggart's paradox

Martin Pleitz

In memory of my teacher Rosemarie Rheinwald (1948–2009).

Usually, the Liar paradox and McTaggart's paradox are treated separately. The first is about *truth*, the second concerns *time*, truth and time being the subjects of distinct subdisciplines of philosophy and philosophical logic. However, as Pavel Tichý's solution of McTaggart's paradox is based on what he calls the "transiency of truth" (Tichý, 1980, p. 165), there may well be connections between truth and time worth exploring. I will argue that those solutions of McTaggart's paradox that relativize the truth of temporally indexical sentences to moments of time (e.g., (Tichý, 1980); (Lowe, 1987)) give a helpful clue for a study of the indexical variant of the Liar paradox, which concerns the following indexical Liar sentence: "This sentence is not true."

1 The indexical fallacy in McTaggart's paradox

McTaggart's paradox starts from two premises:

(P1) Every event is past, present and future.
(P2) Pastness, presentness and futurity are pairwise incompatible.

Together with the uncontroversial further assumption that there is an event these premises imply a contradiction.[1]

The solution is to reject (P1). There are two requirements on this solution: We must *justify* the rejection of (P1) and we ought to *ex-*

---

[1] John McTaggart used this paradox to argue against the reality of *time*, which he took to involve tense essentially (McTaggart, 1908). Nowadays, some philosophers employ the paradox to argue against the reality of *tense* (e.g., (Mellor, 1998, pp. 70ff.)). — As the function of McTaggart's paradox here is only to motivate an indexicalist approach to the Liar paradox, it is not necessary to recount the subsequent debate about the paradox (e.g., (Oaklander & Smith, 1994, pp. 157–285)).

plain the initial plausibility (P1) has had to some philosophers.[2] As Tichý points out, the initial plausibility of (P1) rests on an atemporal reading of the verb "to be". To justify its rejection, one need not go as far as Tichý who holds that all "atemporal predication is illusory" (Tichý, 1980, p. 181). It suffices to note that, while (P2) is unproblematic because it can be construed as a tensed statement that is omnitemporally true, (P1) says something about tensed predicates from a tenseless perspective. Thus, to solve McTaggart's paradox we need only point out that indexical and context-free discourse must not be mixed carelessly.[3]

To avoid the "indexical fallacy" (Lowe, 1987, p. 62), we can follow E. J. Lowe in semantic ascent and give localized truth conditions for tensed sentences. The initial plausibility of (P1) stems from the truth of the following claim: If time is unbounded, then for any temporally bounded event e, there are moments t_1, t_2 and t_3 such that "e is future" is true at t_1, "e is present" is true at t_2 and "e is past" is true at t_3. But given the usual semantics of tensed statements, the moments t_1, t_2 and t_3 must be distinct, so that the conjunction "e is past, present and future" is true at no moment of time (cf. (Lowe, 1987, p, 66)). Therefore we are justified in rejecting (P1).[4]

I will not argue extensively for this particular solution to McTaggart's paradox (which would commit me to deal with the work of those

---

[2] It is a striking historical fact about McTaggart's paradox that although some philosophers have found it obviously fallacious, others have been totally convinced by it. C. D. Broad, though an admirer of McTaggart, calls it a "philosophical howler" (cf. (Oaklander, 1994, p. 157)), while to D. H. Mellor it seems "beyond all reasonable doubt" (Mellor, 1998, p. 72)).

[3] It is important to put emphasis on the word "carelessly" here because indexical and context-free discourse of course *are* often mixed in entirely unproblematic ways. E.g.:
(1) The meeting starts at 12.
(2) Now it is 11:30.
So, (3), we must be on our way now.
While (1) is context-free, (2) and (3) are indexical.

[4] Another example of indexicality misused is what I would like to call the "Sesame Street paradox". Ernie and Bert, who are standing at different places, dispute about who is *here* and who is *there*. To bring out the analogy to McTaggart's paradox, we can describe their situation thus: Ernie is both *here* and *there*, and Bert is both *here* and *there*, but *hereness* and *thereness* are incompatible predicates. The solution is of course that if talk involving the indexical expressions "here" and "there" is relativized to locations in space in the appropriate way, neither Ernie nor Bert will be committed to any inconsistency.

philosophers who have been convinced by it). Rather, I just want to take the following clues from it: The misuse of indexicality can lead to paradox and localizing truth conditions can help to avoid paradox.

2 The Tarskian truth schema and indexical sentences

To move from McTaggart to the Liar, we just have to realize that such context-free truth-conditions for indexical sentences replace the Tarskian truth schema which is essential to the standard derivation of a contradiction from a Liar sentence. As is well known, a definition of truth, according to Alfred Tarski, should imply all sentences that can be obtained from the sentence "s is true if and only if p" by exchanging the symbol "s" for a name of a sentence of the language we are concerned with and the symbol "p" for the translation of that sentence into the language we work with (Tarski, 1956, pp. 187f.) = (Tarski, 1935, pp. 305f.). Ironically, the first instance of the truth schema Tarski gives in *The Concept of Truth in Formalized Languages* concerns a *tensed* sentence:

> "'It is snowing' is a true sentence if and only if it is snowing."
> (Tarski, 1956, p. 156) = (Tarski, 1935, p. 269)[5]

As it stands, this biconditional fallaciously ascribes truth to the sentence-type "It is snowing" just in case it was snowing in 1933 — or just in case it is snowing now (depending on whether we take the present tense of the right hand side to refer to the time of the first publication or to the time of reading). More pressing than this temporal ambiguity of the right hand side is that a quotation expression refers not to a particular token, but to the type of the quoted expression.[6] But it is simply ungrammatical to attribute truth (or falsity) to an indexical sentence[7] *on the type level.*

---

[5] Later, Tarski used a context-free and, in particular, tenseless sentence as his prime example, namely the by now famous "Snow is white" (Tarski, 1944, p. 343).

[6] Tarski himself is quite explicit about the quotation expression in the first instance of the truth schema referring not to a particular token of the quoted expression, but to the class of all equiform tokens; cf. (Tarski, 1956, p. 156) = (Tarski, 1935, p. 269).

[7] I call a sentence *indexical* iff it contains indexical expressions (that are used, not merely mentioned). Thus the truth-value of an indexical sentence will normally, but not in every case, depend on its position of use. E.g., "I am now here", though true at every position of use (here constituted by a triple of a speaker, a moment in time and a place), is indexical.

This has been pointed out as early as 1967 by Donald Davidson. He observes that "the same sentence may at one time or in one mouth be true and at another time or in another mouth be false" and contends that "clearly demonstratives cannot be eliminated from natural language without loss or radical change, so there is no choice but to accommodate theory to them" (Davidson, 1967, pp. 318f.). Davidson accommodates theory by giving local truth conditions like the following:

"'I am tired' is true as (potentially) spoken by p at t if and only if p is tired at t." (Davidson, 1967, pp. 319f.)

So, to account for indexical sentences, the Tarskian truth schema must be substituted by *local* truth conditions, where truth is relativized to the parameter relevant to the indexical expression involved. For instance:

"Today is a snowy day" is true at day d if and only if d is a snowy day.

3 Sentential indexicality: the argument from analogy

Similar local truth conditions will lead to a new perspective on the indexical variant of the Liar paradox. But before turning to the Liar, let me talk about the indexicality of the singular term "this sentence" in general.

In life outside logic, the term "this sentence" often is used to refer to *another* sentence, in a demonstrative or in a context-sensitive way.[8] But used in the indexical Liar sentence it is *purely indexical*, requiring no accompanying gesture and no background information to fix its

---

[8]Some distinctions: An expression is *context-sensitive* iff its reference depends on some feature of its use, where this feature might consist in
(1) some information given in the situation,
(2) an aspect of the situation of use like the speaker or the time of utterance or
(3) an accompanying gesture.
An expression is *co-text-sensitive* iff its reference depends on information given in the situation (1). An expression is *indexical* iff its reference depends in a systematic way on some aspect of the situation the expression is used in (2). An expression is *demonstrative* iff its reference depends on some accompanying gesture (3). Thus indexicals are only one species of context-sensitive expressions. Cf. (Kaplan, 1989, pp. 489–491).

reference. So we can say that by stipulation[9], each token of the term "this sentence" that is used[10] as part of a sentence refers to that sentence. — Now, an expression is indexical if and only if there is a systematic way in which its reference depends on some aspect of the situation the expression is used in. According to Charles Sanders Peirce, who gave an early analysis of indexicality, each token of an indexical expression stands in some "existential connection" to the object it refers to (Burks, 1949, p. 674). Any token of the term "I" refers to the speaker who produced it, any token of the word "today" refers to the day of utterance, etc.

As the term "this sentence" always refers to the sentence it used in, it is a *sententially indexical* term in much the same way the term "I" is personally indexical and the word "today" is temporally indexical. The *existential connection* between each token of the term "this sentence" and the sentence it refers to is given by the relation of parthood.

Because of the indexicality of the term "this sentence", sentences containing it do not fall under the Tarskian truth schema but have local truth conditions. If we let the local truth conditions for personally and temporally indexical sentences be the model, we arrive at the following schema:

"This sentence is Q" is true at sentence s if and only if s is Q.[11]

This gives desirable results for unproblematic sentences. "This sentence consists of six words" comes out as true at itself and as false at "This sentence consists of six words or snow is white" (which in turn comes out as true at itself because of the truth of the second disjunct). So, if a sententially indexical sentence is called "true", this should be seen as elliptical for the attribution of *truth at itself.*

Our willingness to attribute truth simpliciter to a sententially indexical sentence like "This sentence consists of six words" can be ex-

---

[9] Jon Barwise and John Etchemendy make a similar stipulation at the outset of their work on circular propositions; cf. (Barwise & Etchemendy, 1987, p. 16).

[10] If the term "this sentence" is *mentioned*, it will in general *not* refer to the sentence it is a part of. E.g., in "'This sentence' consists of two words", the term "this sentence" does not refer to the sentence it is a part of, while in "This sentence consists of six words", it does.

[11] We might want to restrict the implicit universal quantifier over sentences to those sentences that contain the phrase "this sentence is Q".

plained by the close kinship between a sententially indexical sentence and its *expansion*. The *expansion* of a sententially indexical sentence s is the result of substituting a quotation of s for every occurrence of the term "this sentence" in s.[12] "This sentence consists of six words", has the expansion "'This sentence consists of six words' consists of six words". While the six-word-sentence is indexical and true at itself, its expansion is context-free and true simpliciter.

If a sententially indexical sentence were *synonymous* with its context-free expansion, there would be a reason to identify truth at itself and truth simpliciter. However, even in extension[13], there is a difference between truth at itself and truth simpliciter, as the following pair of a sentence and its expansion shows:

> "This sentence has no token that is embedded in another sentence."

can be uttered and be true at itself, but its expansion

> "'This sentence has no token that is embedded in another sentence' has no token that is embedded in another sentence."

cannot be uttered without being false simpliciter.[14]

---

[12] In particular, if our sentence is of the form "This sentence is Q", then its expansion is of the form "'This sentence is Q' is Q". — The notion of an *expansion* resembles a technical notion Raymond Smullyan (misleadingly) calls "translation": "Let us define the *translation* of a sentence X to be the result of substituting the name of X for every occurrence of 'this sentence' in X." (Smullyan, 1984, p. 203). But there might be differences. It has been argued contra Tarski and Quine that the *quotation* of a sentence is a device of reference of a different kind than a *name* of that sentence because it is possible to (literally) *read off* the sentence from its quotation, but not from a typical name of that sentence (cf. (Reach, 1938); (Anscombe, 1971, pp. 83f.); (Künne, 2007, pp. 180ff.)). And in the context of formal languages we might encounter a *third* kind of device, namely the numeral of an expression's Gödel number.

[13] As they apply to different sentences, truth at itself and truth simpliciter *cannot* have the same extension. So, strictly speaking, what I am arguing against is not the identity of extension, but the claim that for all sententially indexical sentences s, s is true at itself iff the expansion of s is true simpliciter. (Had this claim turned out to be true, we might have been justified in postulating the *synonymy* of each sententially indexical sentence and its expansion. Treating not a sentence but its meaning as the primary bearer of the truth value, this would *in a second step* have justified the thesis that truth at itself and truth simpliciter have the same extension.)

For those of you not taken in by overly complex examples, I want to argue that, even if truth at itself and truth simpliciter were coextensional, at least concerning their *meaning*, there is a difference between a sententially indexical sentence and its context-free expansion. I think that the arguments of Arthur Prior ((Prior, 1959)), Hector-Neri Castañeda ((Castañeda, 1967)), John Perry ((Perry, 1979)) and others for the ineliminability of indexicals show conclusively that an indexical expression adds an own aspect of meaning to the sentence it occurs in. Thus the indexical "This sentence consists of six words" differs in meaning from its context-free expansion "'This sentence consists of six words' consists of six words" just as much as the indexical "I am making a mess", even if uttered by John, differs in meaning from the context-free "John is making a mess" (cf. (Perry, 1979, pp. 12f.)). To put this non-Kaplanian thesis into Kaplanian terms: Not *content* alone, but also *character* is constitutive of meaning (cf. (Kaplan, 1989, pp. 500–507)).

4 Sentential indexicality: the argument from compositionality

The argument for giving local truth conditions to sentences of the form "This sentence is Q" I have just given draws on the analogy of sentential indexicality to other, more mundane cases of indexicality. This argument from analogy can be supplemented by an independent argument from *compositionality*. When we use atomic sentences containing the term "this sentence" and propositional connectives to build non-atomic sentences, strange things seem to happen. Let us look at some examples:

If we apply the Tarskian truth schema to

> "This sentence is negation-free" and to "This sentence is *not* negation-free",

each one comes out as true. So the compositionality of negation seems to be lost.

---

[14] Another example: "This sentence has a token that is used" must be true at itself, but "'This sentence has a token that is used' has a token that is used" can be false simpliciter.

And if we apply the Tarskian truth schema to

"This sentence is atomic" and to "This sentence is not a conjunction",

each one comes out as true. But their conjunction

"This sentence is atomic *and* this sentence is not a conjunction."

comes out as false. So the compositionality of conjunction seems to be lost.[15]

If we apply local truth conditions to the same sentences, we see that compositionality is lost in neither example; rather, it is relativized to positions of use. In the temporal case, we do not count "It is snowing" being true at day one and "It is not snowing" being true at day two as a breach of compositionality. Neither should we see "This sentence is negation-free" being true at sentence one and "This sentence is not negation-free" being true at sentence two as a breach of compositionality!

Davidson tells us with regard to truth relativized to a person and a time that "ordinary logic as now read applies as usual, but only to sets of sentences relativized to the same speaker and time" (Davidson, 1967, p. 319). And similarly, ordinary logic applies as usual to sententially indexical sentences, but only to sets of sentences relativized to the same sentence, conceived as a position of use.

Like the argument from analogy, the argument from compositionality shows that sentences containing the term "this sentence" have local truth conditions. But the reasoning behind it shows more. The

---

[15] These examples for the seemingly non-compositional behavior of sententially indexical sentences are inspired by Smullyan's investigation into what he chooses to call "chameleonic languages", i.e. languages containing a sententially indexical term like "this sentence" (Smullyan, 1984). Though Smullyan notes that the truth set of a chameleonic language will for some sentence contain both the sentence and its negation, but not their conjunction (Smullyan, 1984, pp. 205f. & 212), he does not treat the failure of compositionality in a systematic way. (Smullyan further notes that the truth set of a chameleonic language, though itself contradictory, does not contain *any* contradictory sentence, i.e., no sentence of the form "*p* and not *p*" (Smullyan, 1984, p. 212). But if we extend the concept of sentential indexicality to the subsentential level, the Tarskian truth schema will also deliver true contradictions. E.g., "This subsentence precedes the word 'and' and this subsentence does not precede the word 'and'.")

concept of sentential indexicality requires that a *sentence* can be a position of use in the same way as a speaker or a moment of time. At first glance, this seems weird: A sentence can of course be used by different speakers and at different moments, but what does it mean that a sentence is used *at another sentence*? The idea of composition shows that it *does* make sense to think of a sentence as used at another sentence, namely as used at a longer sentence it is embedded in.[16]

5 The indexical Liar sentence

Now, let us turn to the indexical Liar sentence "This sentence is not true". I will abbreviate its quotation[17] as "λ". If we apply the Tarskian truth schema, we immediately arrive at a contradiction:

$$\lambda \text{ is true if and only if } \lambda \text{ is not true.} \qquad (T_\lambda)$$

But as the term "this sentence" is sententially indexical, we know by now that the Tarskian truth schema must be replaced by a localized truth condition like the following:

$$\lambda \text{ is true at sentence } s \text{ if and only if } s \text{ is not true.} \qquad (L_\lambda)$$

By instantiating the variable "s" as the indexical Liar sentence, we only get this far:

$$\lambda \text{ is true at } \lambda \text{ if and only if } \lambda \text{ is not true.} \qquad (B)$$

Maybe we want to try and be agnostic about the truth of the right hand side of the biconditional (B). We do not know whether λ is true,

---

[16] And accordingly, in the local truth condition

"This sentence is Q" is true at sentence s iff s is Q,

it *does* make sense to instantiate the variable "s" as a sentence different from "This sentence is Q", namely as a longer sentence containing the subsentence "this sentence is Q".

[17] In an investigation of different ways of referring to linguistic expressions, it is important to distinguish quoting and naming, and in particular to distinguish the *abbreviation of a quotation of a sentence* from *a name of that sentence*. λ is only a name of the indexical Liar sentence if a quotation is a name of the quoted expression, which has been disputed (cf. footnote 12). — Note also that λ does not abbreviate the Liar sentence itself, in which case locutions like "λ is true" would have to be spelled out as "This sentence is not true is true" and would thus be ill-formed. I am grateful to Stephen Read for pointing this out to me.

false or neither. Nevertheless we can see that there is no contradiction, because truth at itself and truth simpliciter are different concepts, so that attributing one while negating the other need not be inconsistent. Or maybe we are more daring and say "well, the behavior of λ is so strange that it certainly *cannot be true*", and thus assent to the right hand side of (B). Then we will get the result that λ is indeed *true at itself*. Under the supposition that λ is *not true*, this should be not too surprising — in view of what λ says!

Of course, the mere fact that biconditional (B) is not itself contradictory does not show conclusively that there is no way of deriving a contradiction from the localized truth condition of the indexical Liar sentence (L_λ). But the fact that the standard derivation does not work any longer gives reason to hope that the indexicalist approach to the Liar paradox might lead to its solution.

6   Future work: Liar sentences *without* indexicals

The most pressing problem for the project of an indexicalist solution is constituted by Liar sentences *without* indexicals. Though in many texts the indexical Liar sentence is given as the first example, this is done mostly for heuristic reasons, and the main investigation concerns Liar sentences that achieve the required self-reference by the "more respectable" means of a name or a description.[18] Apart from Epimenides' famous utterance "Everything said be Cretans is not true", there is the Liar sentence "Larry is not true", which is named "Larry", and the Liar sentence "The sentence at position P is not true", which is located at position P (where P might be a *spatiotemporal* position, as in Arthur Prior's Liar sentence ((Prior, 1961, pp. 30f.); cf. (Burge, 1984, p. 94)), or a position *in a printed text*, as in Jan Lukasiewicz' Liar sentence (cf. (Tarski, 1956, pp. 157f.) = (Tarski, 1935, pp. 270f.))).

---

[18]Some examples: (Barwise & Etchemendy, 1987, pp. 3 & 20ff.); (Rheinwald, 1988, pp. 18f.); (Sainsbury, 1995, pp. 111ff.) and (Visser, 2002, pp. 151f.). — A related heuristic use of an indexical term occurs in the informal presentation of a Gödel sentence as similar to a natural language sentences like "I am not provable" or "This sentence is not provable" (cf. (Smullyan, 1984, pp. 201ff.); (Boolos, Burgess, & Jeffrey, 2002, p. 228)). Kurt Gödel himself in his 1931 paper does not give an informal reading of the self-referential formula essential in his proof, but he writes "Wir haben also einen Satz vor uns, der seine eigene Unbeweisbarkeit behauptet." (Gödel, 1931, p. 175)

To justify an extension of the indexicalist solution to cover this plurality of Liar sentences without indexical terms, there are two avenues that I hold to be worth following: Firstly, it should be possible to argue that the extra assumption necessary in reasoning starting from a Liar sentence with a name or a description[19] can be seen as emulating the self-locating belief that is essential in indexical reasoning (cf. (Perry, 1979, pp. 17ff.)). Secondly, the contextual understanding of two-dimensional semantics (cf. (Chalmers, 2006, p. 64)) might be used to show that in a self-referential language, *all* sentences that refer to a sentence behave like indexicals in certain ways.[20]

7 Conclusion

I have shown that because of its sentential indexicality, the indexical Liar sentence must be given local truth conditions. As these take the place of the Tarskian truth schema, the standard derivation of a contradiction is blocked. Though much work remains to be done regarding Liar sentences without indexical terms, I have pointed out that the theory of indexical reasoning and two-dimensional semantics might be used to motivate localized truth conditions for Liar sentences with a name, like Larry, and hopefully also for Liar sentences with a description. Thus there is some reason to expect that an indexical solution can be developed that covers all natural language Liar sentences.[21]

Let me close with a remark about methodology. Anil Gupta and Nuel Belnap give the following methodological advice:

> "When confronted with some perplexing and extraordinary phenomenon, one is tempted to focus on it and to ignore the ordinary and familiar phenomena that are related to it. However, this may not be the best strategy for achieving

---

[19] E.g., the assumption that "Larry" names Larry.

[20] In my presentation at Logica 2009, I gave a sketch of how two-dimensional semantics might be used to show that the peculiarities of a self-referential language lead to names of sentences behaving in an indexical way, following (Tichý, 1984). I have to omit this account here for reasons of space.

[21] I have been careful not to say anything about Liar sentences in *formal* languages. Although I hope that the indexical approach might be further extended to cover an important class of formal Liar sentences, namely those achieving self-reference by means of Gödel numbering, I did not yet want to tackle the difficult question of which referring device of natural language the numerals of Gödel numbers resemble most: names, descriptions, or indexicals.

an understanding of the puzzling phenomenon. [...] Often we come to understand the extraordinary only when we see it in terms of the ordinary. [...] In order to gain a better understanding of the Liar, we need to give *less* attention to the paradoxes than we have given them." (Gupta & Belnap, 1993, p. 17)

I think that, if we give *more* attention to the ordinary uses of indexicals, descriptions and names, and if we are careful in extrapolating from these ordinary uses the more peculiar behavior these devices will show in self-referential languages, we will see the Liar paradox as a *lesser* problem.[22]

Martin Pleitz
Department of Philosophy
University of Münster
Domplatz 23, D-48143 Münster, Germany
martinpleitz@web.de

References

Anscombe, G. (1971). *An introduction to Wittgenstein's Tractatus*. London: St. Augustine's Press.

Barwise, J., & Etchemendy, J. (1987). *The Liar*. New York & Oxford: Oxford University Press.

Boolos, G., Burgess, J., & Jeffrey, R. (2002). *Computability and logic*. New York & Oxford: Oxford University Press.

Burge, T. (1979). Semantical paradox. *Journal of Philosophy, 76*, 169–198.

Burge, T. (1984). Semantical paradox. In R. Martin (Ed.), *Recent essays on truth and the Liar paradox* (pp. 83–117). Oxford: Clarendon Press.

---

[22] I would like to thank Johannes Korbmacher, Stephen Read, Oliver Scholz, Ansgar Seide and Heinrich Wansing for helpful comments on earlier drafts of this text. Special thanks go to Johannes Korbmacher for extensive discussions about this project.

Burks, A. (1949). Icon, index, and symbol. *Philosophy and Phenomenological Research, 9*, 673–689.

Castañeda, H. (1967). Indicators and quasi-indicators. *American Philosophical Quarterly, 4*, 85–100.

Chalmers, D. (2006). Foundations of two-dimensional semantics. In M. García-Carpintero & J. Macià (Eds.), *Two-dimensional semantics* (pp. 55–140). Oxford: Clarendon Press.

Davidson, D. (1967). Truth and meaning. *Synthese, 17*, 304–323.

Gödel, K. (1931). Über formal unentscheidbare Sätze der *Principia mathematica* und verwandter Systeme I. In S. Feferman, J. Dawson, Jr., S. Kleene, G. Moore, R. Soloway, & J. van Heigenoort (Eds.), *Collected works. volume i. publications 1929–1936.* Oxford–New York: Oxford University Press. (1986)

Gupta, A., & Belnap, N. (1993). *The revision theory of truth.* Cambridge: MIT.

Kaplan, D. (1989). Demonstratives. In H. Wettstein, J. Almog, & J. Perry (Eds.), *Themes from Kaplan* (pp. 481–563). Oxford: Oxford University Press.

Künne, W. (2007). *Abstrakte Gegenstände. Semantik und Ontologie.* Frankfurt am Main: Klostermann.

Lowe, E. (1987). The indexical fallacy in McTaggart's proof of the unreality of time. *Mind, 96*, 62–70.

McTaggart, J. (1908). The unreality of time. *Mind, 18*, 457–474.

Mellor, D. (1998). *Real time II.* London & New York: Routledge.

Oaklander, L. (1994). McTaggart's paradox and the tensed theory of time. In *The new theory of time* (pp. 157–162). New Haven–London: Yale University Press.

Oaklander, L., & Smith, Q. (1994). *The new theory of time.* New Haven–London: Yale University Press.

Perry, J. (1979). The problem of the essential indexical. *Noûs, 13*, 3–21.

Prior, A. (1959). Thank goodness that's over. *Philosophy, 34*, 12–17.

Prior, A. (1961). On a family of paradoxes. *Notre Dame Journal of Formal Logic, 2*, 16–32.

Reach, K. (1938). The name relation and the logical antinomies. *Journal of Symbolic Logic, 3*, 97–111.

Rheinwald, R. (1988). *Semantische Paradoxien, Typentheorie und ideale Sprache.* Berlin–New York: de Gruyter.

Sainsbury, R. (1995). *Paradoxes.* Cambridge: Cambridge University Press.

Smullyan. (1984). Chameleonic languages. *Syntese, 60*, 201–224.

Tarski, A. (1935). Der Wahrheitsbegriff in den formalisierten Sprachen. *Studia Philosophica, 1*, 261–405.

Tarski, A. (1944). The semantic conception of truth and the foundations of semantics. *Philosophy and Phenomenological Research, 4*, 341–375.

Tarski, A. (1956). The concept of truth in formalized languages. In *Logic, semantics, metamathematics. Papers from 1923 to 1938* (pp. 152–278). Oxford: Clarendon.

Tichý, P. (1980). The transiency of truth. *Theoria, 46*, 165–182.

Tichý, P. (1984). Kripke on necessity a posteriori. *Philosophical Studies, 43*, 225–241.

Visser, A. (2002). Semantics and the liar paradox. In *Handbook of philosophical logic 11* (pp. 149–240). Dordrecht: Kluwer Academic Publishers.

The Validity Paradox
Stephen Read*

1 The Paradox

An anonymous author of the mid-fourteenth century presents us with a paradox. He asks us to consider the argument:

> God exists
> So this argument is invalid.

Suppose the argument is valid. Then it has a true premise (let us suppose) and a false conclusion. But on a standard account of validity, an argument is valid if and only if it is impossible for the premises to be true and the conclusion false. So any argument with a true premise and a false conclusion is invalid. Hence, on the assumption that the above argument is valid, it follows that it is invalid. So by *reductio ad absurdum*, it is invalid.

However, the proof that it is invalid depends on the assumption, or supposition, that God exists, and that supposition is not only true, but necessarily true, he claims. By a basic principle of modal logic, enunciated by Aristotle in his *Prior Analytics*, whatever follows from a necessary truth is itself necessary.[1] So the conclusion that the argument is invalid is not only true, but necessarily true.

By the standard account of validity, any argument with a necessarily true conclusion is valid. For if the conclusion is necessarily true, it is impossible for the conclusion to be false, and so it is impossible for the premises to be true and the conclusion false. Hence by the standard account of validity, the above argument is valid. But we have proved it is invalid. So the standard account is wrong.

We do not know who the author was of the treatise where this objection to the standard account is given. He is usually referred to

---

*This work is supported by Research Grant AH/F018398/1 (Foundations of Logical Consequence) from the Arts and Humanities Research Council, UK.

[1] *Prior Analytics* 34b22-24: "If, for example, one should indicate the premises by A and the conclusion by B, it not only follows that if A is necessary B is necessary, but also that if A is possible, B is possible."

as "Pseudo-Scotus", since the treatise, a series of questions on Aristotle's *Prior Analytics*, was included in the *Collected Works* of John Duns Scotus in the seventeenth century, but even then was recognised as spurious. The accompanying set of questions on the *Posterior Analytics* is also spurious, and attributed in one manuscript to one "John of Cornwall". However, it seems clear from internal evidence that the two treatises do not have a common author. So our Pseudo-Scotus is not identical with the Pseudo-Scotus who commented on the *Posterior Analytics*, nor with Scotus. But he was aware of the doctrine of the *complexe significabile*, proposed in Oxford in 1331 by Adam Wodeham and taken up in Paris in 1344 by Gregory of Rimini. Other internal evidence suggests a Parisian origin, and so places the work and its author in Paris in the late 1340s or 1350s.

What is most surprising is that Pseudo-Scotus does not seem to recognise the paradoxical nature of his argument. He simply presents it as a refutation of the standard account of validity, proposing to qualify it by adding a clause reading, "except where the meaning of the conclusion is incompatible with the meaning of the inferential sign."[2] But another author writing in Paris in the 1350s, namely, Albert of Saxony, did realise that the argument is paradoxical. For recall Pseudo-Scotus' observation that in proving the argument invalid we supposed only that God exists. What we did was infer the invalidity of the argument from the claim that God exists, that is, we inferred the conclusion of the argument from its premise. But on any account of validity, not just the standard account, any argument whose conclusion follows from its premise is valid. Since the conclusion that the argument is invalid follows from the premise that God exists, the argument really is valid. But it is also invalid, since it has a true premise and false conclusion. Paradox.

Although written in Paris in the early 1350s, Albert's treatise on *Insolubles* is firmly in the English tradition. He was a member of the so-called "English" nation at Paris, though hailing from Saxony in the lower Rhine. As the Hundred Years War between England and France progressed, fewer and fewer English students and scholars went to Paris, and the "English" nation evolved into the "English-German"

---

[2] (Pseudo-Scotus, 2001, p. 228). Cf. (Read, 2001).

The Validity Paradox

nation. Pseudo-Scotus' argument appears as Insoluble XIV in Albert's treatise. Insolubles XI to XIV read:[3]

> XI. 'God exists, and some conjunctive proposition is false' (supposing this is the only conjunctive proposition)
> XII. 'A man is an ass, or some disjunctive proposition is false' (supposing this is the only disjunctive proposition)
> XIII. 'If God exists, some conditional proposition is false' (supposing this is the only conditional proposition)
> XIV. 'God exists, therefore, this consequence is not valid'.

Insoluble XIV results from XIII by replacing a conditional by a consequence, or argument, and 'false' as an epithet of conditionals by the corresponding epithet 'invalid' for arguments. Suppose the conditional in Insoluble XIII is true. Then if God exists, some conditional proposition is false, but by hypothesis, this is the only conditional proposition, so assuming God exists, the conditional is false. That is, if it were true it would be false, so by *reductio ad absurdum*, if God exists, it is false. So it is true, since it says that if God exists it is false, but it is also false, since God does exist (we are supposing). Paradox.

We can run a similar argument with Insolubles XI and XII. Suppose XI is true. Then its second conjunct is true, so, since by hypothesis it is the only conjunctive proposition, it is false, so by *reductio*, it is false. But if it is false, at least one conjunct must be false, and it is not the first conjunct, so the second conjunct must be false, so the conjunction is true. Paradox. Similarly, suppose Insoluble XII is true; then at least one disjunct is true, but not the first disjunct, so the second, so the disjunction is false, so by *reductio* XII is false. So its second disjunct is true, and so XII is true. Paradox.

Suppose we contrapose XIII, then destroy it, so that the only conditional is:

XIII'. 'If some conditional is true then God does not exist'

Now suppose XIII' is true. Then it is a true conditional with a true antecedent, so its consequent is true, that is, God does not exist. That is, if XIII' is true (for, by hypothesis, it is the only conditional), God does not exist. So XIII' is true, since that is what it says, and so God

---
[3](Albert of Saxony, 1988, pp. 357–61).

does not exist, since XIII' is a true conditional with a true antecedent. Now we have not only paradox, but blasphemy.

XIII' is an example of what is nowadays usually called "Curry's paradox", named after Haskell B. Curry. In a paper published in 1942, Curry showed that one could reproduce Russell's paradox in a set theory without negation. Russell's paradox concerns the set of all sets that don't belong to themselves; in symbols, let $R = \{x : x \notin x\}$. We ask if $R \in R$ or not. Suppose $R \in R$, i.e., $R \in \{x : x \notin x\}$. Then $R \notin R$, so by *reductio*, $R \notin R$. It follows that $R \notin \{x : x \notin x\}$, so $R \in R$. Paradox. Curry suggested instead that we look at $C = \{x : x \in x \to p\}$, for an arbitrary proposition p. Then:

$$C \in C \to (C \in \{x : x \in x \to p\})$$
so $\quad C \in C \to (C \in C \to p)$
so $\quad C \in C \to p$
so $\quad C \in \{x : x \in x \to p\}$
so $\quad C \in C$
so $\quad p$.

Curry's paradox leads not just to contradiction, but to triviality, and does so without reference to negation, but apparently relies only on elementary properties of set theory and implication. Curry's paradox was given its natural language version as in XIII' by Geach and Löb, apparently independently, in 1955.[4]

Curry's paradox, and the paradox of validity, constitute a powerful challenge to any theory of truth, of validity, or of sets. If we allow the simple formulation of Insolubles XI–XIV, or the definition of the set C, our theory of truth or of sets is reduced to triviality. For clearly, each of XI–XIV can be adapted to prove any proposition whatever, as in Curry's paradox. For example, we can prove the first disjunct of XII, whatever it is. Suppose XII is false, so its second disjunct is false, so XII is true, so by *reductio* it's true, but its second disjunct is false, so a man is an ass (or whatever). Similarly, we can disprove the first conjunct of XI, the antecedent of XIII, and the premise of XIV. Suppose XIV is valid and its premise is true. Then it's invalid (true premise and false conclusion). So given that its premise is true, its conclusion is true, so it's valid. Hence it's a valid argument with a false conclusion, so its premise is false, that is, there is no God, or whatever. This is a vicious form of paradox. How may it be solved?

---

[4] See (Geach, 1955) and (Löb, 1955).

2 The Solution

Albert of Saxony was a contemporary of John Buridan's at Paris. Buridan's solution to the semantic paradoxes (the insolubles) has been discussed extensively in the past fifty years.[5] Buridan's early suggestion, in, for example, Treatise 7 of his *Summulae de Dialectica*,[6] rested on the claim that every proposition signifies its own truth. So a proposition like the Liar proposition, 'This proposition is false', signifies both that it is false (as it says) and that it is true (like every other proposition), and so is flatly contradictory and hence is simply false. The essential point here is that to be true, everything that a proposition signifies must obtain, in this case, both that it is false and that it is true. This blocks what Maudlin (Maudlin, 2004, p. 8) has recently called Upwards T-Inference (he rejects it too): having shown that the Liar proposition is false, we are not forced to infer that it is (also) true, for not everything which it signifies has been shown to hold. It signifies contradictory things (both that it's true and that it's false), so not everything it signifies could obtain, so it can't be true and is false.

The claim that every proposition signifies its own truth goes back at least to St Bonaventure, writing a hundred years before Buridan.[7] Buridan later came to reject this claim in the last and ninth treatise of his *Summulae* (Buridan, 2001, pp. 968–9), claiming instead that every proposition virtually implies its own truth (that is, if it exists). The reasons for his revision are subtle and depend on Buridan's vehement rejection of the doctrine of *complexe significabilia*; but they need not concern us here.[8] What is more interesting is that in the *Summulae*, Buridan simply asserts that every proposition signifies, or virtually implies, its own truth. He there gives no proof of thes claims. Albert of Saxony, however, adopts Buridan's early theory, and gives a proof, albeit a somewhat simplistic proof, that every proposition signifies its own truth.[9] For every affirmative proposition, he says (roughly), signifies that the subject and predicate stand for the same thing. But for an affirmative proposition to be true is for the subject and predicate

---

[5] See, e.g., (Moody, 1953), (Prior, 1962) and (Hughes, 1982).
[6] See (Buridan, 2001, p. 559).
[7] See (Bonaventure, 1969, pp. 310–311).
[8] See, e.g., (Klima, 2009, ch. 10).
[9] See (Albert of Saxony, 1988, pp. 340–341). Buridan gives a similar argument in (Buridan, 1994, p. 92).

to stand for the same thing. So every affirmative propsition signifies its own truth. Similarly, every negative proposition signifies that the subject and predicate do not stand for the same thing, and that the subject and predicate do not stand for the same thing is what is needed for a negative propositon to be true. So every negative proposition signifies that it is true. Since every proposition is affirmative or negative, every proposition signifies its own truth.

This is hardly a convincing proof that every proposition signifies its own truth. The assumptions on which it is based are too close to the conclusion to avoid the accusation that the argument begs the question. What is more convincing as a ground for a diagnosis of the paradoxes along lines similar to Albert's and Buridan's is the proof Thomas Bradwardine gave of the weaker, but still powerful, thesis that any proposition which signifies its own falsity, or that it is itself not true, also signifies that it is true. Bradwardine's treatise on *Insolubles* was written in Oxford in the early 1320s, and had a strong but largely indirect influence on the approach to the paradoxes throughout the rest of the fourteenth century. In the hundred years before Bradwardine wrote, the prevailing view had been that of the restrictivists (*restringentes*), who variously claimed that no proposition could refer to itself, or to a proposition of which it was a part, but must necessarily refer to some other proposition. So if I say, e.g., that what I am saying is false, I must be referring to my last utterance, or perhaps my next utterance, but not to my present utterance itself. Burley and Ockham, older contemporaries of Bradwardine's, both held this type of view.[10] Bradwardine argued passionately against this and other views before turning to his own.

As with Albert and the early Buridan, a central assumption is that one proposition can signify many things. For example, the proposition 'Socrates is running' signifies not only that Socrates is running, but also that Socrates exists and that someone is running. Indeed, Bradwardine claims that signification is closed under consequence: a proposition signifies every consequence of anything it signifies.[11] This is his second postulate (P2). His first postulate (P1) is Bivalence; every proposition is either true or false, read as including the clause, "and

---

[10]See, e.g., (Spade & Read, 2009, § 2.4).

[11]Actually, what he says is that a proposition signifies everything which follows from it. But in practice he clearly reads that as "everything which follows from what it signifies". Cf. (Spade & Read, 2009, § 3.1).

not both". The third postulate (P3) says that self-reference is possible (against the restrictivists), the fourth (P4) encapsulates the De Morgan principles ("conjunctions and disjunctions with mutually contradictory parts contradict each other"), the fifth (P5) is Disjunctive Syllogism, and the sixth and last (P6) gives the truth-conditions for conjunctions and disjunctions. The two definitions characterise truth and falsity:

(D1) A true proposition is an utterance signifying only as things are

(D2) A false proposition is an utterance signifying other than things are.

Bradwardine's second thesis (T2) is then that every proposition which signifies that it is itself not true or that it is false, also signifies that it is true and is false. Clearly by (P1), and by (D1) and (D2), 'not true' and 'false' are, for Bradwardine, equivalent.

The proof of (T2) is interesting, and set out at length in (Bradwardine, forthcoming, ch. 6). It can be simplified as follows: suppose some proposition, s, signifies that s is not true, and possibly something else as well, call it Q. In symbols, $\mathbf{Sig}(s, \neg\mathbf{Tr}(s) \wedge Q)$. If s were false, then by (D2), something it signifies would fail to obtain, either $\neg\mathbf{Tr}(s)$ or Q, that is, $\mathbf{Fa}(s) \to \neg(\neg\mathbf{Tr}(s) \wedge Q)$, i.e., $\mathbf{Fa}(s) \to (\mathbf{Tr}(s) \vee \neg Q)$. But $\mathbf{Sig}(s, \mathbf{Fa}(s))$, so by (P2), $\mathbf{Sig}(s, \mathbf{Tr}(s) \vee \neg Q)$. Moreover, $\mathbf{Sig}(s, Q)$ and $((\mathbf{Tr}(s) \vee \neg Q) \wedge Q) \to \mathbf{Tr}(s)$, so by (P2) again, $\mathbf{Sig}(s, \mathbf{Tr}(s))$. Hence $\mathbf{Sig}(s, \mathbf{Tr}(s) \wedge \mathbf{Fa}(s))$. By (P1), not both $\mathbf{Tr}(s)$ and $\mathbf{Fa}(s)$, so something s signifies must fail to obtain, whence by (D1), s is false.

What is important here is not so much the proof that s is false as the blocking of Upwards T-Inference. For the paradoxes already establish that propositions such as s are false. The real problem is the apparent implication that they are also true. That move is blocked here, for to use (D1) to show s was true we would have to show that everything it signifies obtains. Given the proof that s signifies its own truth, that is impossible. So the heart of the solution is the proof that any proposition signifying that it itself is false or not true, also signifies that it is true. And that proof is now a substantial and arresting one, drawing this startling conclusion from assumptions which, though powerful, do not obviously beg the question. The key principle is (P2), the closure postulate, that signification is closed under consequence.

In fact, by extending Bradwardine's argument, we can establish Albert's conclusion, that every proposition signifies its own truth. Suppose s signifies Q_1 and Q_2 and so on, in symbols, $\mathbf{Sig}(s, \bigwedge_{i \in I} Q_i)$. Then by (D1),

$$\mathbf{Tr}(s) \Leftrightarrow (\forall p)(\mathbf{Sig}(s,p) \to p) \Leftrightarrow \bigwedge_{i \in I} Q_i$$

so in particular,

$$\bigwedge_{i \in I} Q_i \to \mathbf{Tr}(s).$$

But $\mathbf{Sig}(s, \bigwedge_{i \in I} Q_i)$, so by (P1), $\mathbf{Sig}(s, \mathbf{Tr}(s))$. Thus every proposition signifies that it itself is true.

Given (P2), Bradwardine's analysis solves the standard Liar paradox, and many others.[12] But what of Pseudo-Scotus' validity paradox and the related paradoxes, XI, XII, and XIII? Let us start with XI. Bradwardine discusses it directly (Bradwardine, forthcoming, §8.5):

> "Suppose Socrates utters only this conjunction:
> A: There is a God and Socrates utters a false conjunction, where A_1 is the first conjunct and A_2 the second. Then A is false and A_2 is too, because A_2 signifies that a false conjunction was uttered by Socrates. From this it follows that some conjunction uttered by Socrates is false, and the subject of this conclusion supposits only for A. So this conclusion signifies A to be false, and from 'A is false' it follows that some part of A is false, by (P6),... so either A_1 or A_2 is false... and A_1 is not, so A_2 is, i.e., A_2 is false, so by (P2), A_2 signifies itself to be false."

But the tempting Upwards T-Inference to the conclusion that A is true is blocked, since A_2 signifies not only that A is false but also that A_2 is true, which is not the case.

Bradwardine gives a similar diagnosis of the fallacy in XII (Bradwardine, forthcoming, §8.4):

> "Suppose Socrates utters only this disjunction:
> B: A man is an ass or Socrates utters a false disjunction,

---

[12]See, e.g., (Read, 2006).

and call its second disjunct B_2. Then B is false as is each of its parts, for from B_2 it follows that some disjunction uttered by Socrates is false, but the subject of this conclusion has only one suppositum, namely B, so the conclusion signifies that B is false. From this, by the sixth postulate (P6), it follows that each part of B, and so B_2, is false. So B_2 signifies that B_2 is false. So proceed as before."

Once again, the Upwards T-Inference is blocked, since B_2 signifies not only that B is false but also that B_2 is true, which is not so.

This certainly deals with Insolubles XI and XII. However, it does not yet deal with the general problem raised by Curry's paradox. Many writers on Curry's paradox take 'p' in 'If this proposition is true then p' as a false proposition, just as Bradwardine and Albert take A_1 to be true and B_1 (the first disjunct of B) to be false. Indeed, some even replace 'p' by '$\bot$' (absurdity). But this reduces Curry's paradox to Russell's, given the equivalence of $\neg p$ and $p \to \bot$. What is important, and importantly different, about Curry's paradox (as opposed to the Liar or Russell's paradox) is not what is proved (e.g., a falsehood or a contradiction) as how it is proved. Suppose what replaces 'p' is true, e.g., Fermat's Last Theorem, or that humans have 46 chromosomes. It is still paradoxical that we should be able to prove Fermat's Last Theorem, or this fact about humans, in six lines without any consideration of mathematics or science. Curry's paradox does not just allow us to prove things that are false. It allows us to prove things that are true far too easily.

So what is the diagnosis of Insolubles XI and XII when the first conjunct is false, respectively, true? Take 'p and this conjunction is false', call it A with conjuncts A_1 and A_2. If A is true, then A_2 is true, so A is false, so by *reductio*, A is false. Hence one of the conjuncts is false, but not the second, so the first, so $\neg p$. That last move is too quick. That A is false does not support the conclusion that A_2 is true, since A_2 may signify more than just that A is false, and it may be that further signification of A_2 which means it is false even though A is also false, and indeed, may account for the falsity of A without entailing the falsity of A_1. Suppose A is false. Then if A_2 is true, p is false. So A_2 signifies that if A_2 is true, p is false, by (P2), since A_2 signifies that A is false. So A already involves Curry's paradox. We need first, then to consider Curry's paradox.

Let C be the proposition, 'If C is true then p', and call its antecedent C_1. We quickly establish that if C is true then p. But C_1 signifies that C is true. So C_1 signifies p. So C is true, for it has the (implicit) form 'If C is true and p then p'. But we cannot infer from the fact that C is true that C_1 is true (since the truth of C_1 requires both that C is true and that p), and so it does not follow that C is a true conditional with a true antecedent, so we are not warranted, or required, to infer p.

Returning to A, it follows that A_2 signifies that not-p, so A signifies both that p and that not-p, so A is contradictory and simply false. Similarly, if B is 'Either p or this disjunction is false', B signifies that if B is true then p, so B is a Curry proposition itself, and so is true. Nonetheless, that does not warrant the inference that p, since B_2 signifies not only that B_2 is false (as we saw) but also that not-p, so B is implicitly the tautology, 'Either p or (not-p and B is false)'.

Finally, we must return to Pseudo-Scotus' paradox of validity. Let V be the argument, or inference: 'p, so V is invalid'. Then if V is valid, that V is invalid follows from p, i.e., if V is valid then not-p. That is, if V is valid then not-p, i.e., if p then V is invalid. But this is to infer V's conclusion from its premise, so it is valid, in which case, it's a valid inference with a false conclusion, so its premise is false, so not-p.

The mistake lies in that final step. V is valid, its conclusion is false, and so too is its premise. But it does not follow that not-p. Since V is valid, p entails that V is invalid. Moreover, V's premise signifies that p. So V's premise also signifies that V is invalid, by (P2). So V's premise is false, not because not-p, but because something else it signifies fails to hold, namely, that V is invalid.

3 Conclusion

Paradox can involve not only truth, but also validity, and it can lead not only to contradiction but also to triviality. This is shown by Curry's paradox, capturing the paradoxical nature of Russell's paradox and the Liar paradox without the use of negation, and by Pseudo-Scotus' paradox, which formulates the paradox as an inference rather a proposition. Indeed, we can formulate an inferential version of Curry's paradox as the inference: 'This inference is valid, so p'. We can infer the conclusion from the premise by Modus Ponens, and so the inference is valid, so p. In this way, we could prove anything we liked, so something is sorely amiss.

These paradoxes were familiar fare in the fourteenth century, many of them examined and analysed in Albert of Saxony's treatise on *Insolubles*. Thirty years earlier, Thomas Bradwardine made a novel proposal to deal with them. Any proposition which signifies its own falsehood, he claimed, also signifies its own truth. (In fact, possibly every proposition signifies its own truth, as Albert claimed.) This is a consequence of Bradwardine's second postulate, that signification is closed under consequence. Accordingly, any proposition which signifies its own falsehood, signifies things that cannot all obtain (that is, both that it is true and that it is false) and so is false.

Bradwardine himself showed how to apply his solution to insolubles similar in many ways to Curry's paradox, but only in the degenerate version where p is false, and so expressing little more than the Liar paradox itself. For the really radical version, where p is true, or is an arbitrary proposition, we need to apply Bradwardine's postulate to the proposition afresh. We find that Curry's paradoxical proposition is trivially true, having implicitly the form 'if p then p'. Consequently, it cannot be used to prove p. Similarly, Pseudo-Scotus' paradox has the implicit form, 'God exists and this argument is invalid. So this argument is invalid', trivially valid and no longer warranting the heretical conclusion that there is no God.

Stephen Read
Department of Philosophy, University of St Andrews
St Andrews KY16 9AR, Scotland UK.
slr@st-and.ac.uk
http://www.st-andrews.ac.uk/$\sim$slr/read.html

References

Albert of Saxony. (1988). Insolubles. In N. Kretzmann & E. Stump (Eds.), *Cambridge translations of medieval philosophical texts* (Vol. I Logic and Philosophy of Language, pp. 337–368). Cambridge: Cambridge UP.

Bonaventure. (1969). Is God's existence a truth that cannot be doubted? In J. Wippel & A. Wolter (Eds.), *Medieval philosophy* (pp. 300–313). New York: The Free Press.

Bradwardine, T. (forthcoming). *Insolubilia* (S. Read, Ed.). Leuven: Peeters. (S. Read, Eng. Tr.)

Buridan, J. (1994). *Questiones elencorum* (R. van der Lecq & H. Braakhuis, Eds.). Nijmegen: Artistarium.

Buridan, J. (2001). *Summulae de dialectica*. New Haven: Yale UP. (G. Klima, Eng. Tr.)

Curry, H. (1942). On the inconsistency of certain formal logics. *Journal of Symbolic Logic*, 7, 115–117.

Geach, P. (1955). On *Insolubilia*. *Analysis*, 15, 71–72.

Hughes, G. (1982). *John Buridan on self-reference. Chapter Eight of Buridan's 'Sophismata', translated with an introduction, and a philosophical commentary*. Cambridge: Cambridge UP.

Klima, G. (2009). *John Buridan*. Oxford: Oxford UP.

Löb, M. (1955). On a solution of a problem of Leon Henkin. *Journal of Symbolic Logic*, 20, 115–117.

Maudlin, T. (2004). *Truth and paradox*. Oxford: Oxford UP.

Moody, E. (1953). *Truth and consequence in mediaeval logic*. Amsterdam: North-Holland.

Pironet, F. (1993). John Buridan on the liar paradox: study of an opinion and chronology of the texts. In K. Jacobi (Ed.), *Argumentationstheorie* (pp. 293–300). Leiden: Brill.

Prior, A. (1962). Some problems of self-reference in John Buridan. *Proceedings of the British Academy*, 48, 281–296.

Pseudo-Scotus. (2001). Questions on Aristotle's *Prior Analytics* Q. 10. In M. Yrjönsuuri (Ed.), *Medieval formal logic* (pp. 293–300). Dordrecht: Kluwer.

Read, S. (2001). Self-reference and validity revisited. In M. Yrjönsuuri (Ed.), *Medieval formal logic* (pp. 183–196). Dordrecht: Kluwer.

Read, S. (2006). Symmetry and paradox. *History and Philosophy of Logic*, *27*, 307–318.

Spade, P., & Read, S. (2009). *Insolubles.* Stanford Encyclopedia of Philosophy.

Yrjönsuuri, M. (Ed.). (2001). *Medieval formal logic.* Dordrecht: Kluwer.

Always more

Greg Restall*

A possible world is a *point* in logical space. It plays a dual role with respect to propositions.

(i) A possible world determines the truth value of every proposition. For each world w and proposition p, either at w, p is true, or at w, p is not true.

(ii) Each set of possible worlds determines a proposition. If $S \subseteq W$ is a set of worlds, there is a proposition p true at exactly the worlds in S.

Perhaps such a proposition is not expressible in any language that you or I speak, but – so a familiar story goes – it is decided by each world, so it plays just the role that other propositions do, so it counts as a proposition in the same way. In fact, we can see just how it counts as a proposition: given all the worlds in S, our proposition p says that the world is one of the worlds in S. It describes a way the world is, even if we have no means of picking out the set S, so it is a proposition.[1]

But does this talk of possible worlds actually make sense?

Metaphysical worries about worlds are well known. These worries do not concern the role they play in the analysis of propositions: they call into question the 'otherness' of worlds, the profligacy of admitting locales where there are tailless kangaroos or blue swans. Worries of this sort can be assuaged by giving an account of worlds which takes them to be abstract, or fictions, or in some other way less real than the world you and I are thought to inhabit. Ontological profligacy is not so much of a concern if we have understood worlds in a metaphysically

---

*Thanks to the Logic Seminar at the University of Melbourne (especially Allen Hazen and Lloyd Humberstone), Shawn Standefer, and the audience at LOG-ICA2009 for discussions on these matters.
This research is supported by the Australian Research Council, through grant DP0343388, and Sam Phillips' *Don't Do Anything*.

[1] David Lewis' *On the Plurality of Worlds*, (Lewis, 1986, ch. 1), has a very good defence of this position on the relationship between worlds and propositions, but the view is not just his. The view is everywhere.

thin manner. The fact that this concern is so easily sidestepped shows that this concern is does not touch (i) and (ii) – the properly logical notion of a possible world. In this paper, I will to consider the logical structure of commitment to (i) and (ii). Do claims such as (i) and (ii) have any unforeseen *logical* costs?

* * *

It would seem like there is little reason to reject (i) and (ii). To transpose talk of worlds into an algebraic key, structures satisfying (i) and (ii) are well known. They are complete atomic Boolean algebras. In such algebras, atoms play the role of possible worlds: at each atom, the propositions *entailed* by that atom can be taken to be true, and the others are false. The fact that the algebra is complete means that every collection of atoms determines a proposition in the appropriate way: any set of atoms has a least upper bound, which is true at those and only those atoms. Atomicity gives us (i) and completeness gives us (ii). If there are reasons to reject the combination of (i) and (ii), then the construction of complete atomic Boolean algebras must somehow not apply.

In this paper, I will construct a logic, extending classical logic with a single unary operator, which has *no* complete Boolean algebras as models. If the family of *propositions* we are talking about in (i) and (ii) has the kind of structure described in that logic, then (i) and (ii) cannot jointly hold.

* * *

The new operator, $\#$, may be introduced in a straightforward manner. Here is the first cut at an account of $\#$. Take a propositional language with the infinite supply of atoms $p_1, p_2, p_3, \ldots$ and define '$\#A$' to be the first propositional atom not occurring in the formula A.[2]

$\#$ has interesting logical properties. Since $\#A$ is an atom not occurring in A, if A is satisfiable, so is $\#A \wedge A$, and so is $\neg\#A \wedge A$. In fact, if we can derive A from $\#A$, then A is a tautology. Similarly,

---

[2] In other words, for now, $\#p_1$ *is* the atom p_2. So, $\#\#p_1 = \#p_2 = p_1$, even though there is a sense that 'p_1' does 'occur in' $\#\#p_1$). That is not the relevant sense here.

if $\#A$ is derivable from A, then A itself is unsatisfiable. We have the following four principles[3]

$$\begin{array}{ll} \text{If } \#A \vdash A \text{ then } \vdash A. & \text{If } \vdash A, \#A \text{ then } \vdash A. \\ \text{If } A \vdash \#A \text{ then } A \vdash. & \text{If } A, \#A \vdash \text{ then } A \vdash. \end{array} \qquad (\#)$$

Now, as defined, $\#A$ is not anything like a connective: it is a syntactic device. It is not a congruence with respect to logical equivalence, since $\#p_1 = p_2$ but $\#(p_1 \wedge (p_1 \vee p_2)) = p_3$, even though p_1 is logically equivalent to $p_1 \wedge (p_1 \vee p_2)$.

We can remedy this by setting $\#A$ to be defined as the first propositional atom which is not in *some* formula equivalent to A. Then this satisfies substitutivity of equivalents. Now, $\#(p_1 \wedge (p_1 \vee p_2)) = p_2$, since there is some formula equivalent to $p_1 \wedge (p_1 \vee p_2)$ (namely, p_1)) in which p_2 doesn't occur, but there is *no* formula equivalent to $p_1 \wedge (p_1 \vee p_2)$ in which p_1 doesn't occur. It is straightforward to verify that $\#$ so defined still satisfies the four conditions given in $(\#)$.

* * *

Now, consider the logic extending classical propositional logic with an operator $\#$ satisfying the four $(\#)$ conditions.[4] A logic of this form can have well-defined models. We have seen one, with $\#$ defined syntactically. Logics extending classical logic with $\#$ make sense, and are coherent. There is nothing inconsistent or incoherent in the logic of $\#$.

However, the logic is still *odd*. While a logic like this can have a Boolean algebra as a model — the Lindenbaum algebra of equivalence classes of provably equivalent formulas will do as an example — they have no atomic Boolean algebras as models. Recall: a is an *atom* in a Boolean algebra if for every element x either $a \wedge x = 0$ (the bottom

---

[3] Where, as usual, we take $X \vdash Y$ to hold if and only if there is no evaluation where each member X is true and each member of Y is not. So, $X \vdash$ when X cannot all be true together, and $\vdash Y$ when Y cannot all be false together.

[4] I will not call them *rules* for they are not inference rules *defining* the connective $\#$. A language can contain two independent operators $\#_1$ and $\#_2$ both satisfying the conditions $(\#)$. In fact, one way to understand Kaplan's paradox over the size of the collection of possible worlds is to think of 'I believe that p' as $\#p$. For it seems that whether I believe p or not is genuinely logically independent from p, at least when p is logically contingent (Kaplan, 1995). I owe this observation to Allen Hazen.

element in the algebra) or $a \wedge x = a$. There are no elements between 0 and a. An algebra is said to be *atomic* if every element is the join of some collection of atoms.

Now, since every finite Boolean algebra is atomic, every model for the logic will be *infinite*. But not every infinite Boolean algebra will work, either. The algebra of all subsets of some infinite set — ordered by inclusion, and with the usual Boolean operators of intersection, union and complementation — will not do either, since each singleton set is an atom.

Here is why no algebra for $\#$ is atomic. Take a Boolean algebra with an atom a. Consider $\#a$. By the conditions $(\#)$, since the atom a is neither 1 nor 0 (it is neither a tautology nor a contradiction) then neither $a \wedge \#a$ nor $a \vee \#a$ are tautologies nor contradictions. But this is inconsistent with a's being an atom, for $a \wedge \#a$ entails a but is not the bottom element of the algebra. So, the algebra is not only not atomic, but it cannot contain *any* atoms.

This talk about *algebras* and *atoms* has consequences for theories of *worlds*. (i) and (ii) commit us to taking the collection of propositions to be an atomic Boolean algebra. If each proposition is modelled by the set of worlds in which it is true, and if *every* set of worlds models a proposition, then each *singleton* set of worlds is an atom. It is true *somewhere*, but there is no non-trivial proposition stronger than it. This rules out $\#$, or it rules out taking (i) and (ii) to jointly hold.

* * *

Perhaps $\#$ is a mere syntactic device. Is it an artefact of the presentation of a language with infinitely many atomic sentences? Can we specify a model for a $\#$ satisfying the $(\#)$ conditions in which the language of sentences plays no special role? I will explain here how we can construct just such models, to provide a language-invariant structure in which the Boolean connectives and $\#$ may be interpreted. Here's how.

Let's think in terms of worlds, to start. Take the set W of *worlds* to be the irrational numbers in the Real Line. The propositions at *Level* n are the unions of the any selection of irrational intervals of length $\frac{1}{2^n}$: $(\frac{z}{2^n}, \frac{z+1}{2^n})$ where z is an integer. These are closed under union (the union of any collection of intervals is a collection of intervals), intersection (there is no worry about endpoints of abutting intervals, as these don't reach their endpoints, which are rational) and complement (the complement of some collection of intervals is the collection of the other intervals: since the endpoints are rational, they don't occur in either a set or its complement). The propositions at each level are *finer* classifications of points than at any of the previous level.

Propositions at Level 0 are collections of intervals such as $(-2, -1)$, $(0, 1)$, $(3, 4)$, etc. Propositions at Level 1 are collections of *finer* intervals $(-1.5, -1)$, $(1.5, 2)$, $(3, 3.5)$, etc., and so on, throughout each finite level.[5]

Let's interpret sentences in the language of propositional logic — enhanced with the operator '#' — as propositions at *some* level or other. If A and B are interpreted as propositions at Level n, then $\neg A$, $A \wedge B$ and $A \vee B$ are also interpreted as propositions at Level n, since the union, intersection or complement of propositions at Level n are also at Level n.

To interpret $\#A$, where A is interpreted as a proposition at Level n, we will choose a proposition at Level $n+1$. In particular, we will choose an *alternating* proposition at Level $n+1$: the proposition consisting of all of the intervals $(\frac{z}{2^n}, \frac{z+1}{2^n})$ where z is even integer. The alternating proposition at Level 0 is

$$\cdots (-4, -3) \cup (-2, -1) \cup (0, 1) \cup (2, 3) \cup (4, 5) \cdots$$

the alternating proposition at Level 1 is

$$\cdots (-2, -1.5) \cup (-1, -0.5) \cup (0, 0.5) \cup (1, 1.5) \cup (2, 2.5) \cdots$$

and so on. This choice for $\#A$ satisfies the four conditions given in $\#$. Let A be interpreted as a proposition at Level n. If A is not true everywhere, then it is false at some interval $(\frac{z}{2^n}, \frac{z+1}{2^n})$. Now consider $\#A$. It is true at $(\frac{2z}{2^{n+1}}, \frac{2z+1}{2^{n+1}})$, where A is not true. And $\#A$ is not true at $(\frac{2z+1}{2^{n+1}}, \frac{2z+2}{2^{n+1}})$, where A is not true. Similarly, if A is not false everywhere, it is true at some interval $(\frac{z}{2^n}, \frac{z+1}{2^n})$. $\#A$ is true at $(\frac{2z}{2^{n+1}}, \frac{2z+1}{2^{n+1}})$, where A is true. And $\#A$ is not true at $(\frac{2z+1}{2^{n+1}}, \frac{2z+2}{2^{n+1}})$, where A is true. In other words, $\#A$ is truly independent of A. If A is true somewhere, at some such places, $\#A$ is true, and at others, $\#A$ is false. If A is

---

[5]We could just as easily do this with regions on a grid in irrational 2-space, or cubes in 3-space, etc. We have all the generality we need in one dimension, however.

false somewhere, at some such places, $\#A$ is true, and at others, $\#A$ is false.

This fact is completely general. For any proposition A we have found another proposition $\#A$. $\#A$ is more finely grained than A, and the four rules of extensibility are satisfied. In models like these, it makes sense to think of '$\#$' as an operator on propositions, and not merely a syntactic device for constructing sentences from other sentences. The language may now be finite, or indeed it may have *no* non-logical constants![6]

So, we have a syntax-free model in which the four conditions ($\#$) hold, so we must have either of (i) and (ii) failing if we are to think of these points as *worlds*. It is easy to see which. Not every set of worlds is a proposition. Only some sets of points — those at some Level or other — count as a proposition. Others are not.

However, we do not need to think of this model in that way. We could, instead, take (i) to fail, if we wish to avoid commitment to worlds altogether. The appeal to worlds in these models is not essential: we could instead refrain from all talk of worlds and appeal instead to *regions* in a formal topological space. The definition of propositions in terms of sets of points — irrational numbers in our case — is not essential. The construction gives us an atomless Boolean algebra, and these are well known algebraic structures. The value of the relatively concrete construction here is the manner in which extensibility corresponds to propositions being more and more finely grained, without that ever coming to an end. The model shows that the idea of indefinite extensibility of propositions is coherent: and operators like $\#$ are one way to give formal structure to the intuitive idea that the collection of propositions is indefinitely extensible. Wherever we find ourselves in the collection of propositions, we haven't exhausted its depths. For any proposition at all, there is always more.

Greg Restall
Philosophy Department
The University of Melbourne

---

[6] If there are no non-logical constants, but the logical constants $\bot$ and $\top$, then we can still construct the alternating propositions at each level, and a whole host of other propositions. Consider, for example, what $\#\top \wedge \#\#\top$ and $\#\top \wedge \neg\#\#\top$ are, to get a feel for what propositions may be constructed.

restall@unimelb.edu.au
http://consequently.org/

References

Kaplan, D. (1995). A problem in possible world semantics. In W. Sinnott-Armstrong, D. Raffman, & N. Asher (Eds.), *Modality, morality and belief: Essays in honor of ruth barcan marcus* (pp. 41–52). Cambridge: Cambridge University Press.

Lewis, D. K. (1986). *On the plurality of worlds.* Oxford: Blackwell.

Evaluation Games for Shapiro's Logic of Vagueness in Context

Christoph Roschger*

1 Introduction

Stewart Shapiro (Shapiro, 2006) presents a model for reasoning with vague propositions with a special focus on Sorites situations (Hyde, 2008). He maintains that the extensions and anti-extensions of vague predicates such as *bald* and *red* strongly depend on the conversational context. At the beginning of a conversation this context is empty; the extensions and anti-extensions of vague predicates are undefined for many objects, the so-called *borderline cases*. During a conversation these notions are sharpened, such that borderline cases, which have been undecided so far, get assigned to the (anti-)extension of the vague predicates in question.(It is the counterpart to the notion of *supertruth* in supervaluationist theories) Shapiro introduces logical connectives operating on formulas containing such vague predicates. Additionally to the classical connectives, he introduces new ones operating globally on trees of possible contexts.

This contribution introduces a Hintikka-style game for evaluating formulas according to Shapiro's model of vagueness. This is motivated by the following two observations:

- Shapiro's main setting, a so-called forced march version of the Sorites paradox (Hyde, 2008), already includes dialogue situations and conversational records. A dialogue game to evaluate composite propositions is just a natural consolidation of this concept.

- The game provides an explicit mechanism for the evaluation of formulas. In particular Shapiro's falsehood and indefiniteness conditions for global connectives and quantifiers are rather indirect. The dialogue rules provide a much more direct and me-

---

*This work is supported by Eurocores-ESF/FWF grant 1143-G15 (LogICCC-LoMoReVI).

chanical characterization of truth in a model. As we will see, the defined connectives can be expressed in terms of a finite two-player zero-sum game with perfect information.

2 Shapiro's approach to vagueness

Below I sketch the main points of Shapiro's account of vagueness as presented in (Shapiro, 2006) and (Shapiro, 2008). This sketch is in no way complete; topics irrelevant for the dialogue game are left out. For example, Shapiro's treatments of higher order vagueness and vague objects within his framework are not subject of the presented game, and thus are omitted here.

Central notions of Shapiro's work are *judgment dependence*, *open texture*, and the *principle of tolerance*. *Judgment dependence* means that the extensions and anti-extensions for the borderline cases of vague predicates are solely determined by the decisions of competent speakers. Vagueness in Shapiro's framework is characterized by judgment dependence. More generally, Shapiro holds that the extension and anti-extension of a vague predicate uttered in a conversation depend on the conversational context. Decisions made by the conversationalists are put on the *conversational record* together with (explicit or implicit) assumptions. This includes, for example, assumptions, statements made by them so far, and (logical) consequences thereof. Moreover it is possible that statements can, explicitly or implicitly, be withdrawn from the conversational record, which plays a crucial role when in Sorites situations.

Open texture means that for a vague predicate P there exists an object a such that a competent speaker can decide that $P(a)$ holds or that $P(a)$ does not hold without her competency being compromised. Note that the notion of 'competent speaker (of the English language)' is also vague; this is where the model can be extended to higher order vagueness.

The *principle of tolerance* is closely related to open texture. Its precise formulation as used in (Shapiro, 2008) is:

> Suppose that two objects a, b in the field of P differ only marginally in the relevant respect (on which P is tolerant). Then if one competently judges a to have P, then she cannot competently judge b in any other manner.

Games for Shapiro's Logic

The main settings described by Shapiro are so-called forced march Sorites situations: Imagine 2000 men lined up where man #1 has full hair and man #2000 has no hair at all. The men are ordered by their amount of hair. A group of conversationalists is repeatedly asked if they judge man #i as bald, starting with man #1, continuing until man #2000 is reached. At each step we require them to return a communal verdict. At the beginning, they will unequivocally vote for 'not bald', but at some point they will begin to discuss and finally switch to 'bald'. Shapiro holds, that at this point not only information is added to the conversational record, but also the last few judgments are implicitly retracted.

2.1 Shapiro's model theory

In order to reflect these notions in the model theory, Shapiro uses a Kripke-like tree structure, called *frame*. Each frame, denoted $\langle W, M \rangle$, consists of a set of worlds W with one designated world $M \in W$ called the *base* of the frame. A world is a partial valuation of atoms assigning either *true*, *false*, or *indefinite* to all predicates in question[1]; all worlds in a frame are over the same domain. The world N' is called a sharpening of the world N, denoted as $N' \succeq N$, if and only if each atom which is *true* or *false* at N is also *true* or *false*, respectively, at N'. At the base M propositions are fixed which are determined outside the current conversation. This includes (non-linguistic) facts, external contextual factors, relevant thoughts and practices, etc. Thus it is required that for all $N \in W$, the world N is a sharpening of M. As $\succeq$ is a partial order, a frame can be considered as a tree of precisifications with root M. Note that, in contrast to supervaluationist approaches (Fine, 1975), the completability requirement is not enforced. This means that we do not require that at the leaves of the tree structure a vague predicate P is decided for all objects, i.e., we may leave P undecided for some objects. Shapiro argues that in Sorites situations complete sharpenings (where P is decided for all instances) violate the principle of tolerance (or the externally determined facts that man #1 is not bald and man #2000 is bald), thus they are artifacts of the model theory.

---

[1] Shapiro defends the notion that there are conceptually only two truth values, *true* and *false*; *indefinite* is to be interpreted as the absence of a classical truth value.

In a Sorites situation initially only the externally determined facts are available on the conversational record. Making (competent) judgments corresponds to moving alongside a branch, away from the root M and thus precifying the asserted statements. In the beginning of a forced situation the conversationalists will repeatedly vote for 'not bald' until, at some point they will switch to 'bald'. With the principle of tolerance in force they have to withdraw some statements from the conversational record; this amounts to jumping to another branch in the frame. It is possible to formalize from which worlds to which worlds such jumps are allowed and where not.

Shapiro argues that *determinate truth* in a frame is best characterized by the notion of *forcing*. A formula ϕ is forced at a sharpening N, if for each sharpening N' of N there is a further sharpening N'' of N' such that ϕ holds at N''. Intuitively, ϕ being forced at N means that ϕ will eventually get *true*: a formula ϕ is *determinately true* at if ϕ is forced at the base of F. (*Determinate truth* is the counterpart to the notion of *supertruth* in supervaluationist theories.) Moreover, the notions of validity and, more generally, consequence are defined in terms of forcing: $\Gamma \models \phi$ if and only if ϕ is forced at every sharpening in every frame in which all formulas of $\Gamma = \{\psi_1, \psi_2, \ldots\}$ are forced.

Of course not all possible frames are adequate for a given (Sorites) situation. For example, we can exclude frames which contain partial interpretations where man $\#i$ is declared to be bald, but another man $\#j$ with $j > i$, who has more hair, is judged not to be bald. Such constraints on adequate frames are called *penumbral connections*. They do not have to check each sharpening separately; it is also possible to require that some condition holds not locally at a sharpening but globally for the frame, e.g. by requiring that some proposition is forced at the base. Note that tolerance can also be formulated as a penumbral connection.

2.2 Defining connectives and quantifiers

2.2.1 Local operators

Shapiro first defines local logical connectives for negation '$\neg$', conjunction '$\wedge$', disjunction '$\vee$', and implication '$\rightarrow$'. These all adhere to the standard Kleene truth tables as given by Figure 1, where 0 denotes *false*, 1 denotes *true*, and u denotes *indefinite*. The quantifiers '∃' and '∀' are defined as expected.

∧	0	u	1		∨	0	u	1		→	0	u	1		¬	
0	0	0	0		0	0	u	1		0	1	u	0		0	1
u	0	u	u		u	u	u	1		u	1	u	u		u	u
1	0	u	1		1	1	1	1		1	1	1	1		1	0

Figure 1: Kleene truth tables for local connectives

Note that all these connectives obey to the monotonicity principle on sharpenings which we have encountered above for atomic propositions. That means, if a (compound) formula ϕ is *true* (or *false*) at a sharpening N then ϕ is *true* (or *false*, respectively) also at all sharpenings N' of N. Because of this, forcing is not present in the object language, but only at the meta-level: it is easy to construct a frame where it is *false* that a formula ϕ is forced at a sharpening N, but where it is *true* that ϕ is forced at a sharpening N' of N.

2.2.2 Global operators

Additionally to the standard logical connectives and quantifiers, Shapiro introduces new non-local ones operating on whole subtrees instead of a single sharpening. One of them is the new non-local implication '⇒' with the following semantics:

> $\phi \Rightarrow \psi$ is *true* at a sharpening N if at each sharpening N' of N if ϕ is *true*, then also ψ is *true*.

This connective is used extensively by Shapiro to define penumbral connections as seen by the following example: assume a Sorites situation as explained above. Then we can stipulate as a penumbral connection, that for all i and j the formulas $(B(m_j) \wedge S(m_i, m_j)) \Rightarrow B(m_i)$ and $(\neg B(m_i) \wedge S(m_i, m_j)) \Rightarrow \neg B(m_j)$, with $S(x, y)$ iff x has more hair than y, hold at the base (and thus at all sharpenings). This ensures that at a sharpening where man m_i is judged 'bald', all men with less hair are judged 'bald' as well. Vice versa, at a sharpening where man m_j is judged 'not bald', all men with more hair as m_j are judged 'not bald' as well.

In order to preserve monotonicity we also give a falsehood condition for each new connective. Just stating that $\phi \Rightarrow \psi$ is *false* if it is not *true* would violate monotonicity; this can be seen in the following example frame:

```
              M
            /   \
  N : P(a),¬P(b)   N' : P(a),P(b)
```

At the base M the formula $P(a) \Rightarrow P(b)$ is not *true* because the condition is violated at N, but it is *true* at N'. Thus, the only way not to violate the monotonicity principle is to leave $P(a) \Rightarrow P(b)$ undecided at M. Therefore, Shapiro makes use of the so-called *stable failure*:

> The formula $P(a) \Rightarrow P(b)$ is *false* at the sharpening N if and only if there is no sharpening N' of N such that $P(a) \Rightarrow P(b)$ is *true* at N.

This ensures that also falsehood is preserved in the tree structure, thus if a formula $\phi \Rightarrow \Psi$ is *false* at a sharpening N, it is also *false* at each sharpening of N.

Another new connective is the intuitionistic-style negation '−'. The proposition $-P(a)$ is *true* at the sharpening N if there is no sharpening N' of N where $P(a)$ is *true*.[2]

Shapiro observes that, as in supervaluationist theories, a formula $\exists x.\phi(x)$ can be forced at a sharpening N without $\phi(a)$ being forced at N for any particular witness a. In order to make the existence of such witnesses expressible in the object language, he introduces a new global existential quantifier E with the following semantics:

> The formula $Ex.\phi(x)$ is *true* at N if and only if there exists a such that $\phi(a)$ is forced at N.

Similarly it is possible to define the new global universal quantifier A:

> The formula $Ax.\phi(x)$ is *true* at N if and only if for all x it holds that $\phi(x)$ is forced at N.

---

[2] Notice that a formula ϕ is forced at a sharpening N if and only if $\neg -\phi$ is *true* at N. However, the property of being forced cannot be introduced as an unary connective at the object level. This still violates monotonicity as is not the case that $\neg -\phi$ is *false* if and only if ϕ is *not* forced at N.

As seen above, it is also necessary to give falsehood conditions for all the new connectives in order to preserve monotonicity. Therefore the falsehood conditions for '$-$', 'E', and 'A' are obtained by their stable failure analogously to '$\Rightarrow$'.

We obtain the following lemma:

Lemma 1. *Let ϕ be a formula of the form $-\psi$ or $Ax.\psi(x)$: ϕ is indefinite at a sharpening N in F if and only if there exist sharpenings N' and N'' of N such that ϕ is true at N' and false at N''.*

Proof. If ϕ is *true* at N' and *false* at N'' then, due to monotonicity, it can neither be *false* nor *true* at N. Therefore it must be *indefinite* at N. On the other hand, consider, e.g., $\phi = Ax.\psi(x)$, and assume that ϕ is *indefinite* at N. According to the definition of stable failure there exists at least one sharpening of N where ϕ is *true* (otherwise ϕ would be *false* at N). Assume that there exists no sharpening where ϕ is *false*. Then ϕ is either *true* or *indefinite* at any given sharpening N' of N. But, as just argued, if ϕ is *indefinite* at N' there exists a further sharpening N'' where ϕ is *true*. This means that ϕ is forced at N. Shapiro shows that a formula $Ax.\psi(x)$ is forced at a sharpening exactly if it is *true* at that sharpening, and consequently we conclude that ϕ is *true* at N leading to the desired contradiction. Thus there exists a sharpening N'' of N such that ϕ is *false* at N''. For the global negation '$-$' we can reason analogously.

Notice that Lemma 1 does not hold for formulas of the form $\phi \Rightarrow \psi$ or $Ex.\phi(x)$. Such formulas can be *indefinite* at a sharpening N and *true* at all further sharpenings of N.

3 A Hintikka-style evaluation game

3.1 Motivation and Overview

As we have seen above, Shapiro's logic directly refers to conversational situations, namely a forced march version of the Sorites paradox, but involves only atomic predicates. A dialogue game to decide the semantic status of a compound formula without leaving this dialogue setting therefore just seems a natural consolidation of this concept. Moreover, the game provides an explicit mechanism for the evaluation of formulas. All moves consist of either choosing between different alternatives how the game should proceed, selecting a representative of the domain, or selecting a sharpening of the current one. As we will see, especially falsehood and indefiniteness conditions for global operators are specified much more directly this way.

There are two players, the proponent **P** of a formula and the opponent **O**. Initially **P** asserts that a formula ϕ is either *true*, *false*, or *indefinite* at an initial sharpening N in a given frame F. This is denoted as **P** asserting $\vdash_N^+ \phi$, $\vdash_N^- \phi$, or $\vdash_N^\sim \phi$, respectively. During the game ϕ is decomposed step by step into less complex formulas according to the game rules until, in the end, **P** asserts the semantic status of only an atomic formula $P(a)$. Assume that the game ends at the sharpening N'. Then, if at this point **P** asserts $\vdash_{N'}^+ P(a)$ and if a is in the extension of P at N' then **P** is declared the winner of the game, otherwise **P** loses and **O** wins. Analogously, **P** wins if he asserts $\vdash_{N'}^- P(a)$ and a is in the anti-extension of P at N', or if he asserts $\vdash_{N'}^\sim P(a)$ and a is neither in the extension nor in the anti-extension of P at N'.

Both players are assumed to agree on the frame and the initial sharpening in the frame in which to evaluate the formula. The game is a finite two-player zero-sum game with perfect information. Thus, by Zermelo's Theorem (Zermelo, 1912) we conclude that the game is determined.

3.2 Dialogue Rules

As described above, at each point in the game exactly one formula is asserted by the proponent **P** to be *true*, *false* or *indefinite* at a certain sharpening. The dialogue rules then specify how this formula is to be further reduced and which player has to make which choices based on the outmost connective or quantifier. For instance, the dialogue rule for conjunction is given in Figure 2. It can be read as follows: if **P** asserts $\vdash_N^+ \phi \wedge \psi$ than **O** can choose whether **P** has to further assert $\vdash_N^+ \phi$ or $\vdash_N^+ \psi$ at the same sharpening. For $\vdash_N^- \phi \wedge \psi$, on the other hand, **P** himself may choose. If **P** asserts $\vdash_N^\sim \phi \wedge \psi$ he first chooses whether to assert that both ϕ and ψ are *indefinite* at N, or that only ϕ is *indefinite* and ψ is *true*, or vice versa. In response **O** chooses one of the two corresponding assertions.

As one can see, the rules can be obtained directly from the Kleene truth tables in Figure 1. Informally, $\phi \wedge \psi$ is *indefinite* at a sharpening

Games for Shapiro's Logic

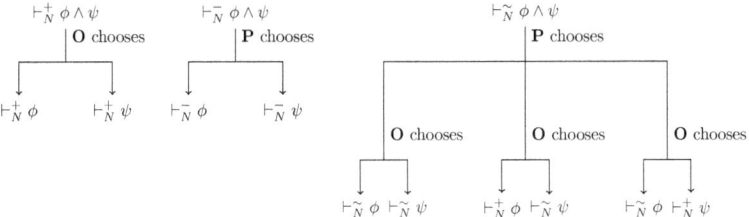

Figure 2: Dialogue rule for conjunction

N, if either ϕ is *true* and ψ is *indefinite*, or vice versa, or both are *indefinite* at N. Rules for the other local connectives can be constructed analogously.

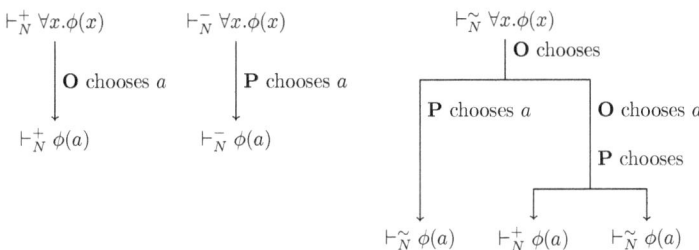

Figure 3: Dialogue rule for the local universal quantifier

For local quantifiers we proceed in the same way: Figure 3 shows the dialogue rules for the universal quantifier. For $\vdash_N^+ \forall x.\phi(x)$ **O** has to choose one domain element a and the game proceeds, whereas for $\vdash_N^- \forall x.\phi(x)$ the choice is **P**'s. In the third case, $\vdash_N^\sim \forall x.\phi(x)$, first **O** chooses whether he wants **P** to select one element a and assert $\vdash_N^\sim a$ or if he wants to select a by himself, but let **P** choose whether to assert $\vdash_N^+ \phi(a)$ or $\vdash_N^\sim \phi(a)$. This rule can be informally motivated by observing that $\forall x.\phi(x)$ is *indefinite* if and only if for all instances a of x it holds that $\phi(a)$ is either *true* or *indefinite* and, moreover, there is at least one instance a' such that $\phi(a')$ is *indefinite*. Again, a dialogue rule for the existential quantifier can be obtained analogously.

The dialogue rules above are all *local* in the sense that the current sharpening is not changed. The rules for the other, global, connectives involve choosing sharpenings of the current one by **P** or **O**.

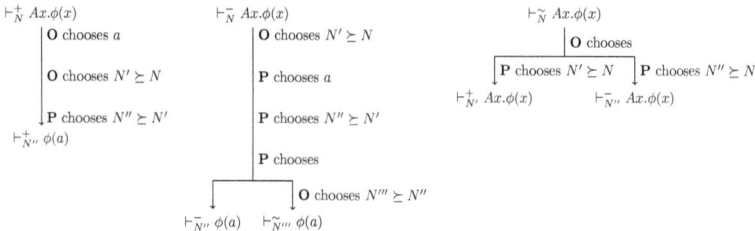

Figure 4: Dialogue rule for the global universal quantifier

The rule for the global universal quantifier 'A' is given in Figure 4. In the rule for $\vdash_N^+ Ax.\phi(x)$ first **O** selects a domain element a and then chooses a sharpening N' of N. Then **P** chooses yet a further sharpening N'' of N' and asserts $\vdash_{N''}^+ \phi(a)$. According to Shapiro's definition, in order for $Ax.\phi(x)$ to be *true* at N, after **O** has chosen a, the formula $Ax.\phi(x)$ must be forced at N. By letting players alternatively select further sharpenings we obtain a literal translation of Shapiro's forcing condition to dialogue rules. The rule $\vdash_N^- Ax.\phi(x)$ involves Shapiro's definition of the stable failure of 'A'. According to this definition **P** has to show that there is no sharpening of N where $Ax.\phi(x)$ is *true*. Thus, after **O** chooses $N' \succeq N$, player **P** selects a domain element a and then shows that $\phi(a)$ is not forced at N'. This is the case, if he can find a sharpening $N'' \succeq N'$ where either $\phi(a)$ is *false*, or $\phi(a)$ is *indefinite* and remains so in all further sharpenings. The rule $\vdash_N^{\sim} Ax.\phi(x)$ is directly obtained from Lemma 1 stating that $Ax.\phi(x)$ is *indefinite* at N if and only if there exists sharpenings N' and N'' such that $Ax.\phi(x)$ is *true* at N' and *false* at N''.

The dialogue rules in Figure 5 for the global negation '$-$' follow the same scheme: Rules for $\vdash_N^+ -\phi$ and $\vdash_N^- -\phi$ are obtained directly from Shapiro's truth and falsehood conditions; the rule for $\vdash_N^{\sim} -\phi$ is again obtained from Lemma 1. Since, as noted above, forcing can be

Games for Shapiro's Logic 241

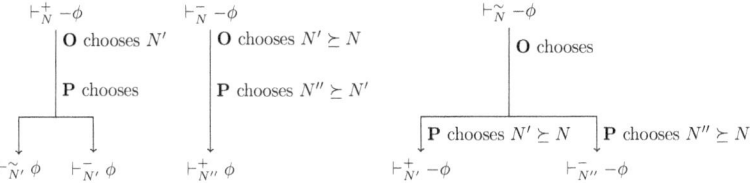

Figure 5: Dialogue rule for the global negation

expressed in terms of this operator, we can read the dialogue rule for $\vdash^{-}_{N} -\phi$ directly as a rule for forcing: if the proponent **P** wants to state that a formula ϕ is forced at a sharpening N, he does so by asserting $\vdash^{-}_{N} -\phi$.

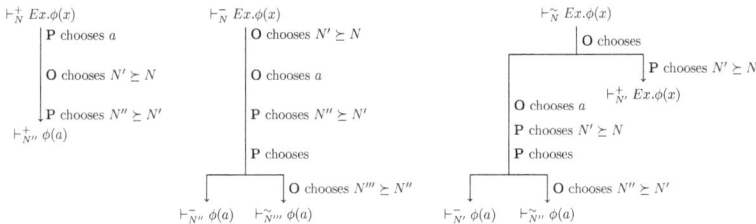

Figure 6: Dialogue rule for the global existential quantifier

Figure 6 shows the dialogue rules for the global existential quantifier 'E'. The difference between the rules for $\vdash^{+}_{N} Ex.\phi(x)$ and $\vdash^{-}_{N} Ex.\phi(x)$ and their counterparts for the global universal quantifier is only that here the proponent **P** has to pick one element of the domain instead of the opponent **O**. However, as Lemma 1 does not hold for the global existential quantifier 'E', we have to give another rule for $\vdash^{\sim}_{N} Ex.\phi(x)$: if **P** asserts that $Ex.\phi(x)$ is *indefinite* at N, he has to be able to show that it is not *true* and to show that it is not *false* at N. The former case amounts to the left branch: for any given element a, player **P** asserts that $\phi(a)$ is not forced by providing a sharpening N' of N where $\phi(a)$ is either *false*, or *indefinite* and remains *indefinite* at each further sharpening. In the latter case **P** shows that the $Ex.\phi(x)$ is not *false* at N by providing a sharpening of N where the formula is *true*.

The connective for the global implication '$\Rightarrow$' is used by Shapiro solely to formulate penumbral connections, that are constraints on possible frames. As the game is an evaluation game which takes place in a given frame, such contraints are not directly subject to the game. However, one can still specify dialogue rules for the '$\Rightarrow$' connective in the same way as for the others. Due to space restrictions the exact formulation of these rules is omitted here.

3.3 Adequacy of the game

We claim that the dialogue rules are adequate for Shapiro's logic in the following sense:

Theorem 1. *Given a frame F and a sharpening N in F, a formula ϕ is true at N in F if and only if the player* **P** *has a winning strategy for the game where he initially asserts $\vdash^+_N \phi$. ϕ is false at N if and only if* **P** *has a winning strategy for the game where he initially asserts $\vdash^-_N \phi$ and* indefinite *if and only if* **P** *has a winning strategy for the game where he initially asserts $\vdash^\sim_N \phi$.*

Proof. We proof by induction on the complexity of ϕ that the game rules are adequate for Shapiro's logic. If ϕ is atomic, this is obvious. Otherwise, applying one of the dialogue rules reduces ϕ to a less complex formula except for the rules for $\vdash^\sim_N Ax.\phi(x)$, $\vdash^\sim_N Ex.\phi(x)$, and $\vdash^\sim_N -\phi$. However, the latter cases reduce to the respective rules for $\vdash^+_N Ax.\phi(x)$, $\vdash^-_N Ax.\phi(x)$, $\vdash^+_N Ex.\phi(x)$, $\vdash^+_N -\phi$, and $\vdash^-_N -\phi$ and therefore are covered by the induction as well. Due to space restrictions this checking of rules is only carried out here for some exemplary ones.

Assume, for example, that **P** asserts $\vdash^+_N \forall x.\psi(x)$. If $\forall x.\psi(x)$ is *true* at N, then no matter which domain element a player **O** chooses, $\psi(a)$ is *true* at N. By the induction hypothesis player **P** asserting $\vdash^+_N \psi(a)$ wins the game. On the other hand, if $\vdash^+_N \phi$ is not *true* at N, then there exists an element b such that $\psi(b)$ is not *true* at N. If **O** selects b, player **P** has to assert $\psi(b)$, and, again by applying the induction hypothesis we see that **P** loses the game.

As a slightly more complex example assume that **P** asserts $\vdash^\sim_N Ex.\psi(x)$. Player **P** wins if and only he can show that $Ex.\psi(x)$ is neither *true* nor *false* at N. Assume, **O** chooses the left branch. Then **P** wins if for each domain element a he can find a sharpening N' such that either $\psi(a)$ is *false* at N' or there for all sharpenings of N' it holds that $\psi(a)$ is *indefinite* at N'. But this exactly means that there is a sharpening of N where there is no further sharpening such that $\psi(a)$ is *true*; in short, $\psi(a)$ is not forced at N. Since a was chosen by **O**, player **P** wins if there is no a which is forced at N. In other words, $Ex.\psi(x)$ is not *true* at N. On the other hand, if **O** chooses the right branch, **P** wins if there exists a sharpening of N where $Ex.\psi(x)$ is *true*, thus, according to the definition of stable failure, **P** wins when $Ex.\psi(x)$ is not *false*. Since **O** chooses between the left and the right branch, it is the case that **P** wins exactly if $Ex.\psi(x)$ is neither *true* nor *false* at N.

As noted above, forcing can be expressed in terms of the global negation '−'. The following corollary follows immediately from Theorem 1.

Corollary 1. *Given a frame F and a sharpening N in F, a formula ϕ is forced at N in F if and only if the player **P** has a winning strategy for the game starting in $\vdash^-_N -\phi$.*

4 Conclusion and future work

In this contribution we have presented a dialogue game for the evaluation of formulas in Shapiro's logic in a given frame. At each point in the game the initial proponent of the formula ϕ in question asserts that a subformula of ϕ is *true*, *false*, or *indefinite* at a given sharpening. Compound formulas are being subsequently reduced to less complex formulas until, in the end, an atomic formula can easily be evaluated by checking the (anti-)extensions of the vague predicates in question at the final sharpening reached. The dialogue rules consist of simple operations like choosing a domain element, choosing a sharpening, or choosing between different succeeding assertions, which yields a rather mechanic characterisation for the connectives of Shapiro's logic.

In future work we plan to investigate other types of games adequate for this logic. In particular evaluation games in the spirit of dialogue games as defined by Paul Lorenzen (Lorenzen, 1960) and, more specifically, by Robin Giles (Giles, 1974) for Łukasiewicz logic seem promising. Such games strictly separate the stepwise decomposition of compound formulas into their atomic parts from the evaluation of atomic game states. In contrast to the game presented here, both players may assert a multiset of formulas at each point in the game. The characterisation of indefiniteness is an interesting property of this game: we can observe that *truth* of a formula ϕ coincides with the existence of a winning strategy for the player asserting ϕ in the beginning, while *falsehood* coincides with the existence of a winning strategy for the other player. For indefinite formulas neither player has a winning strategy, which fits Shapiro's point of view that indefiniteness in his logic is not just a third truth value, but merely signifies the lack of a classical one.

Christoph Roschger
Theory and Logic Group, Vienna University of Technology
Favoritenstr. 9-11/ E1852, A-1040 Wien, Austria
roschger@logic.at
http://www.logic.at/staff/roschger/

References

Fine, K. (1975). Vagueness, truth and logic. *Synthese*, *30*(3), 265–300.

Giles, R. (1974). A non-classical logic for physics. *Studia Logica*, *33*(4), 397–415.

Giles, R. (1976). Lukasiewicz logic and fuzzy set theory. *International Journal of Man-Machine Studies*, *8*, 313–327.

Hintikka, J., & Sandu, G. (1996). Game-theoretic semantics. In J. van Benthem et al. (Ed.), *Handbook of Logic and Language*. Elsevier.

Hyde, D. (2008). Sorites paradox. In E. N. Zalta (Ed.), *The Stanford Encyclopedia of Philosophy* (Fall 2008 ed.).

Lorenzen, P. (1960). Logik und Agon. In *Atti del congresso internazionale di filosofia* (pp. 187–194). Firenze: Sansoni.

Shapiro, S. (2006). *Vagueness in context*. Oxford University Press.

Shapiro, S. (2008). Reasoning with slippery predicates. *Studia Logica*, *90*(3).

Zermelo, E. (1912). Über eine Anwendung der Mengenlehre auf die Theorie des Schachspiels. In *Proceedings of the Fifth International Congress of Mathematicians* (Vol. 2, pp. 501–10). Cambridge.

A New Notion of Meaning Connection and the Logic of Simple Processes

Igor Sedlár and Juraj Podroužek*

1 Introduction

The presence of the so-called paradoxes of material implication is commonly taken as evidence of the fact that the formulas with material implication as their main connective are an inappropriate formal analysis of natural language conditionals. It is often argued that at least some natural language conditionals have to be analysed in terms of some sort of meaning connection between the antecedent and the consequent and, of course, the definition of material implication does not involve reference to any kind of meaning connection whatsoever. More precisely, semantics of the classical propositional logic is too weak to represent any reasonable notion of meaning and thus too weak to represent any kind of meaning connection.

Needless to point out, the attempts to construct logics with meaning-sensitive implication connectives have been an important part of research in philosophical logic since the beginning of 20th century. It is crucial to observe, however, that any attempt to formulate a logical calculus containing a meaning-connection-sensitive implication connective has to be preceded by a more or less explicit pre-theoretical specification of the intended meaning connection. According to C. I. Lewis, the desired meaning connection is analysable in terms of necessity or impossibility:[1]

---

*Work on this paper was supported by the grant No. 1/0814/08 of the Slovak Ministry of Education and Slovak Academy of Science (VEGA), and by the grant UK 2009/101 of Comenius University.
This work was supported by the grant No. 1/0162/09 of the Slovak Ministry of Education and Slovak Academy of Science (VEGA). Both authors would like to express their gratitude to J. Szomolányi and M. Zouhar for their helpful remarks on earlier drafts of this paper, and to the audience at Logica '09 for their interesting comments.
[1]Cf. mainly (Lewis & Langford, 1932).

(S) 'A implies B' is true iff it is impossible that 'A' is true and 'B' is false.

Lewis' modal calculi are attempts to make (S) formally precise.

Now, the paradoxes of strict implication are commonly taken as evidence that necessity is not enough and the story continues: the meaning connection based solely on necessity is not adequate and so a stronger pre-theoretical idea has to be formalized to obtain more adequate propositional logics. And so, a new family of logics is born out of the idea of relevance:[2]

(R) 'A implies B' is true iff there is a justification of 'B' that depends solely on the presupposition of 'A' and 'A' has to be really used in the justification.

Are there any other reasonable kinds of meaning connection? If so, what logical calculi arise on their ground? The aim of this paper is to provide partial answers to these questions. We formulate what we believe to be a new notion of meaning connection between the antecedent and the consequent of a true conditional. After that we will sketch a logical system based on the new notion, namely the Logic of Simple Processes (LSP). We will investigate into the relations of LSP to other logics with nonstandard implication connectives. Then we will outline the interpretation of LSP as a logic of justifications, possibly suitable for representing knowledge as justified true belief. We will conclude by mentioning the most important open problems.

2 Meaning connection via verification procedures

Consider the following statements:

If a natural number x is divisible by 6, then it is divisible by 2. (1)

If every natural number divisible by 6 is even,

then π is irrational. (2)

Now, (1) is true and we certainly tend to claim that there is some kind of meaning connection between its antecedent and its consequent.

---
[2]Cf. mainly (Anderson & Belnap, 1975), (Anderson, Belnap, & Dunn, 1992), (Mares, 2004).

What kind of connection is it? Sure, it is not possible for a natural number to be divisible by 6 and not to be divisible by 2 and so (1) satisfies (S). But so does (2), assuming that 'π is irrational' is necessarily true. It is plausible to claim however, that the connection between the antecedent and consequent is more strict in (1) than in (2). After all, what has the divisibility by 6 and 2 to do with the irrationality of π? Moreover, it is certainly not true that for every number x divisible by 2 there is a justification of the statement 'x is divisible by 2' that depends solely on the assumption that x is divisible by 6 and that really uses this assumption, for there are numbers divisible by 2 but not divisible by 6. So, the meaning connection in statement (1) is not the connection described by (R). So, again, what kind of meaning connection are we dealing here with?

The answer is easy for everyone who remembers the simple divisibility-by-6 algorithm taught in elementary schools: to demonstrate that a given number is divisible by 6, demonstrate that it is divisible by 2 and 3. Thus, dealing with every number a divisible by 6, we verify the truth of the statement 'a is divisible by 6' by verifying the truth of 'a is divisible by 2' and then 'a is divisible by 3' or the other way around. More generally, in the context of the simple algorithm, every successful verification procedure of 'x is divisible by 6' contains a successful verification procedure of 'x is divisible by 2'. Of course, it is this easy only if we confine ourselves to the simple divisibility-by-6 algorithm. But the idea is clear.

Now consider the following principle:

(P) 'A implies B' is true iff every admissible verification procedure of 'A' contains an admissible verification procedure of 'B'.

It is plain that this principle presupposes at least two important things. First, we presuppose that it is possible to determine which procedures are verification procedures of a given statement. Of course, we do not confine ourselves to *actual* procedures, for in this case we would be able to consider true statements only. We only demand that it is possible to describe what *would have to be done* in order to demonstrate that a given statement is true. In other words, we are dealing with *hypothetical* verification procedures. Moreover, (P) requires the possibility to select some of the possible procedures of a given statement as 'admissible'. This only means that we do not have to consider every possible procedure for a given statement, but we may concentrate only on some of them.

Second, (P) is specifying a kind of meaning connection between statements, and so it presupposes a link between the meaning of a statement and possible verification procedures of the statement. We assume that there is a very close connection between understanding a statement's meaning and the ability to describe what would count as a verification of the statement.[3]

3 The logic of simple processes

The present section contains a sketch of a logical system based on the ideas mentioned above. It is the Logic of Simple Processes (hereafter LSP). First, it is necessary to make the idea of verification procedures more precise. We take for granted that verification procedures are sequences of basic actions, or steps. We do this by defining the following two sets:

Definition 1. The set of *atomic steps* is the countable set $At = \{\mathfrak{e}, a_1, a_2, \ldots\}$.
The set of *simple processes* over At is the set $Proc = \{\langle a_i, \ldots, a_j \rangle \mid \{a_i, \ldots, a_j\} \subseteq At\}$.

The members of $Proc$ are finite tuples of members of At. We take the members of $Proc$ to be formal representations of verification procedures and we name them 'simple processes'. $Proc$ is the set of finite simple processes over At. Observe that the set At is, besides its cardinality, unspecified. Informally, we take this set to be a formal representation of basic steps (or atomic steps) within verification procedures. The notion of a basic step is somewhat vague and this is also a reason why we take members of At to be undefined basic blocks of (the semantics of) LSP. The 'empty' process $\mathfrak{e}$ is a special member of At and we discuss it in more detail later.

Let A, B, etc., be variables ranging over the members of $Proc$ and X, Y, etc., variables ranging over nonempty subsets of $Proc$. Let $A(i)$ be the ith member of A, similarly for $B, \ldots, j, \ldots$.

Definition 2. $A \sqsubseteq B$ iff $A(k) = B(n_k)$ where $n_1 < n_2 < \ldots$ is an increasing sequence of indices ('*A is a subprocess of B*').

---

[3](P) also presupposes that it is possible for a procedure to be contained in another procedure. We assume that this is not problematic.

Definition 3. $X \supseteq_P Y$ iff $\forall A \in X \exists B \in Y (B \sqsubseteq A)$ ('X P-includes Y'). $A \sqsubseteq B$ makes precise the idea of one procedure being contained in another procedure. P-inclusion is a notion directly leading to a formal representation of (P): X P-includes Y iff every simple process from X contains a simple process from Y as its subprocess.

Lemma 1. $\supseteq_P$ is

(a) *reflexive, transitive, but*

(b) *not antisymmetric, and*

(c) *not symmetric.*

Proof. (a) follows immediately from the definition of $\supseteq_P$. For (b) consider sets $\{\langle c \rangle\}$ and $\{\langle a,c \rangle, \langle c \rangle\}$. For (c) consider sets $\{\langle a,b,c \rangle\}$ and $\{\langle a,b \rangle\}$.

Now we introduce the propositional logic of simple processes LSP. Its language $\mathcal{L}_{\mathsf{LSP}}$ contains of a countable set of propositional variables $\{p_1, p_2, \ldots\}$, a propositional constant $\bot$, and the connetctives $\neg$, $\wedge$, $\vee$, $\rightarrow$. The set of formulas of $\mathcal{L}_{\mathsf{LSP}}$ is defined in the usual way. The set of literals Lit is the set of propositional variables and their negations.

Next, we define the valuation function:

Definition 4. A valuation v_i is a function from Lit to the set of singletons over $Proc$, $v_i : Lit \rightarrow \{\{A_j\} | A_j \in Proc\}$, satisfying the condition: if $v_i(p) = \{A_k\}$ and $v_i(\neg p) = \{A_l\}$, then $A_k \neq A_l$.

Every valuation v_i yields a special subset of $Proc$, the set of 'inadmissible processes' $\bot_i$:

Definition 5. If $v_i(p) = \{A\}$ and $v_i(\neg p) = \{B\}$, then $A \circ B \in \bot_i$ and $B \circ A \in \bot_i$. If $C \in \bot_i$ and $C \sqsubseteq D$, then $D \in \bot_i$ where '$A \circ B$' denotes the concatenation of processes A and B.

One more definition:

Definition 6. Let X, Y, and Z be sets of simple processes. Z is *a minimal antecedent set of Y with respect to X* (a m. a. s.) iff

- $(Z \circ X) \supseteq_P Y$ and there is no Z' such that $Z \supseteq_P Z'$ and $(Z' \circ X) \supseteq_P Y$, or

- $(X \circ Z) \supseteq_P Y$ and there is no Z' such that $Z \supseteq_P Z'$ and $(X \circ Z') \supseteq_P Y$.

Every valuation can be extended to an interpretation:

Definition 7. An interpretation is a function $\mathcal{I}$ from the set of formulas to the set of nonempty subsets of *Proc* satisfying the following conditions:

- If $\varphi \in Lit$, then $\mathcal{I}(\varphi) = v_i(\varphi)$;
- $\mathcal{I}(\bot) = \bot_i$;
- If $\mathcal{I}(\varphi) = X$ and $\mathcal{I}(\psi) = Y$, then $\mathcal{I}(\varphi \wedge \psi) = X \circ Y$;
- If $\mathcal{I}(\varphi) = X$ and $\mathcal{I}(\psi) = Y$, then $\mathcal{I}(\varphi \vee \psi) = X \cup Y$;
- If $\mathcal{I}(\varphi) = X$ and $\mathcal{I}(\psi) = Y$, then

$$\mathcal{I}(\varphi \to \psi) = \begin{cases} \{\mathfrak{e}\} & \text{if } X \supseteq_P Y, \\ \bigcup Z_i : Z_i \text{ is a m. a. s. of } Y \text{ wrt } X & \text{if } X \not\supseteq_P Y. \end{cases}$$

- If $\varphi \notin Lit$, then $\mathcal{I}(\neg\varphi) = \mathcal{I}(\varphi \to \bot)$

It is plain that our conjunction is not commutative, because of the non-commutativity of concatenation. The main motivation behind our understanding of implication is intuitionistic. We define $\mathcal{I}(\varphi \to \psi)$ as the union of minimal sets that need to be concatenated with $\mathcal{I}(\varphi)$ so that the result will P-include $\mathcal{I}(\psi)$. But what if $\mathcal{I}(\varphi)$ already P-incudes $\mathcal{I}(\psi)$? The intuitive answer would be to 'do nothing'. This is the role of the special 'empty' process $\mathfrak{e}$.

Definition 8. A formula φ *holds* under an interpretation $\mathcal{I}$ iff $\{\mathfrak{e}\} \supseteq_P \mathcal{I}(\varphi)$ (in symbols: $\mathcal{I} \models \varphi$). A formula φ is *valid* iff it holds under every interpretation (in symbols: $\models \varphi$). A formula ψ is a P-*consequence* of φ iff $\mathcal{I}(\varphi)$ P-includes $\mathcal{I}(\psi)$ for every $\mathcal{I}$ (in symbols: $\varphi \supseteq_P \psi$).

It is natural to assume that if a formula holds under some interpretation, then it is valid, for every interpretation gives the same empty process, so to speak. This is, however, not the case.

4 LSP-validity

This section contains some preliminary results on comparing LSP with other logics and on some interesting properties of LSP itself.

Lemma 2. *If $\varphi \supseteq_P \psi$, then $\models \varphi \to \psi$.*

Proof. If ψ is a P-conseqence of φ, then $\mathcal{I}(\varphi) \supseteq_P \mathcal{I}(\psi)$ for every $\mathcal{I}$. But then, by the definition of $\mathcal{I}(\varphi \to \psi)$, $\mathcal{I}(\varphi \to \psi) = \{\mathfrak{e}\}$, for every $\mathcal{I}$. But this means that $\varphi \to \psi$ holds under every interpretation, hence $\models \varphi \to \psi$.

Note that the converse statement is not true (cf. Theorem 1.(f)).

Theorem 1. *The following schemata and formulae are LSP-valid:*

(a) $\varphi \to \varphi$;

(b) $(\varphi \to \varphi) \to (\psi \to \psi)$;

(c) $(\varphi \wedge \psi) \to \varphi$;

(d) $\varphi \to (\varphi \vee \psi)$;

(e) $\neg(p \wedge \neg p)$;

(f) $p \to (q \to q)$.

Proof. (a) follows from Lemma 1. (b) follows from (a), the definition of $\mathcal{I}(\varphi \to \psi)$, and Lemma 2. (c) and (d) are consequences of definitions of $\mathcal{I}(\varphi \wedge \psi)$, $\mathcal{I}(\varphi \vee \psi)$, and of Lemma 2. For (e) note, that $\mathcal{I}(\neg(p \wedge \neg p)) = \mathcal{I}((p \wedge \neg p) \to \bot)$ and that $(p \wedge \neg p) \supseteq_P \bot$. For (f) note that $p \to (q \to q)$ holds in every interpretation. Consider the two possible cases: (i) $\mathcal{I}(p) \supseteq_P \{\mathfrak{e}\}$. Then $\mathcal{I}(p \to (q \to q)) = \{\mathfrak{e}\}$ by (a) and the definition of $\mathcal{I}(\varphi \to \psi)$. (ii) $\mathcal{I}(p) \not\supseteq_P \{\mathfrak{e}\}$. Then $\{\mathfrak{e}\} \in \mathcal{I}(p \to (q \to q))$ and thus $\mathcal{I} \models p \to (q \to q)$. So, $p \to (q \to q)$ holds in every interpretation and is valid.

Observe that (b) and (f) of the previous theorem yield that LSP does not have the variable sharing property and thus is stronger than the mainstream relevant logics. Is it equivalent to classical logic, intuitionistic logic, or some modal logic based on strict implication?

Theorem 2. *The following schemata and formulae are not LSP-valid:*

(a) $p \vee \neg p$;

(b) $q \to (p \vee \neg p)$;

(c) $(p \wedge \neg p) \to \varphi$.

Proof. For (a) consider any interpretation that does not make $\mathcal{I}(p) = \{\mathfrak{e}\}$, or $\mathcal{I}(\neg p) = \{\mathfrak{e}\}$. For (b) consider an interpretation where $\mathcal{I}(q) = \{A\}$, $\mathcal{I}(p) = \{A \circ B\}$, and $\mathcal{I}(\neg p) = \{C\}$ provided that $\mathfrak{e} \not\sqsubseteq B$ and $\mathfrak{e} \not\sqsubseteq C$. Then $\mathcal{I}(p \to (q \vee \neg q)) = \{B, C\}$ and hence the formula is not valid in $\mathcal{I}$. For (c) consider an interpretation, where $\mathcal{I}(p) = \{A\}$, $\mathcal{I}(\neg p) = \{B\}$, and let φ be such that $\mathcal{I}(\varphi) = \{A \circ B \circ C\}$, provided $\mathfrak{e} \not\sqsubseteq C$. Then $\mathcal{I}((p \wedge \neg p) \to \varphi) = \{C\}$, and hence the formula does not hold in the interpretation.

Item (a) proves that LSP is not equivalent to classical logic and gets our logic closer to intuitionistic logic. Item (c), however, proves that LSP is not equivalent to intuitionistic logic, neither to any modal logic that validates *ex falso quodlibet*.

5 A logic of justifications?

The formal simple-process semantics of LSP and its informal inepretation in terms of the hypothetical verification procedures give rise to the following questions. Is it possible to think of LSP as a logic of justifications? What is the relation of LSP to the Artemov-style[4] justification logics? And, in general: Is it possible to augment LSP in a way that would yield a logic of knowledge as justified true belief? What follows is a brief sketch of attempts in this direction.

First, we may introduce a set of agents $Ag = \{\alpha_1, \alpha_2, \ldots, \beta, \ldots\}$ and a set of states $St = \{s_1, s_2, \ldots\}$. Now we may think of agents as actually performing certain verification procedures in certain states. For example, agent α_1 performed the procedure $\langle a, b, c \rangle$ in state s_1 and the procedure $\langle a, b, c, d \rangle$ in state s_2. In general, we may define a verification function $\mathcal{V} : Ag \times St \longrightarrow Proc \cup \{\emptyset\}$. The function $\mathcal{V}$ assigns to every state the procedures performed by the agents in that state.[5]

---

[4]Cf. mainly (Artemov, 2001), (Artemov, Kazakov, & Shapiro, 1999), (Fitting, 2004), (Fitting, 2005), (Fitting, 2009).

[5]If $\mathcal{V}(\alpha, s) = \emptyset$, then α is idle in s: α performs no procedures. This must not be confused with performing the special empty procedure $\mathfrak{e}$.

Logic of Simple Processes

We may presuppose that if some agent performs a procedure A in some s_i and $A \in \mathcal{I}(\varphi)$ for some formula φ, then φ is true in s_i with respect to the given interpretation.[6] Moreover, if an agent α performs A, then α justifies the truth of φ, and thus has a *justified* true belief of φ. In other words, α knows φ in s with respect to the given interpretation. But it is possible to claim that α has verified so much more! Namely, α has verified every formula ψ such that $\mathcal{I}(\varphi) \supseteq_P \mathcal{I}(\psi)$.

We may sum up these ideas in the following tentative definition of knowledge in LSP:

Definition 9. Let $\alpha \in Ag$, $s \in St$, and φ is a formula. Then α *knows* φ in s with respect to the tuple $\langle \mathcal{I}, \mathcal{V} \rangle$ iff $\mathcal{V}(\alpha, s) \supseteq_P \mathcal{I}(\varphi)$ (in symbols: $(\mathcal{I}, \mathcal{V}, s) \models K_\alpha \varphi$).

This notion of knowledge has interesting properties:[7]

Theorem 3. *The following hold:*

(a) *If* $\varphi \supseteq_P \psi$, *then* $(\mathcal{I}, \mathcal{V}, s) \models K_\alpha \varphi$ *implies* $(\mathcal{I}, \mathcal{V}, s) \models K_\alpha \psi$ *for every* α, s, $\mathcal{V}$ *and* $\mathcal{I}$.

(b) *It is not generally true that* $\models \varphi$ *implies* $(\mathcal{I}, \mathcal{V}, s) \models K_\alpha \varphi$ *for every* α, s, $\mathcal{V}$ *and* $\mathcal{I}$.

(c) *It is not generally true, that if* $\mathcal{I} \models \varphi$ *and* $\mathcal{I} \models \psi$ *and* $(\mathcal{I}, \mathcal{V}, s) \models K_\alpha \varphi$, *then* $(\mathcal{I}, \mathcal{V}, s) \models K_\alpha \psi$.

Proof. (a) follows from the definition of knowledge, the definition of P-inclusion and from Theorem 1. Let $\mathcal{V}(\alpha, s) = \{A\}$ and $\{A\} \supseteq_P \mathcal{I}(\varphi)$ (α knows φ) and $\varphi \supseteq_P \psi$. Then $\{A\} \supseteq_P \mathcal{I}(\psi)$ due to Lemma 1. and hence α knows ψ.

For (b) note that it is not generally true that $\mathfrak{e} \sqsubseteq \mathcal{V}(\alpha, s)$, or the fact that it is possible, that $\mathcal{V}(\alpha, s) = \emptyset$.

For (c) consider the case where $\mathcal{I}(\varphi) = \{\langle a, b \rangle, \mathfrak{e}\}$, $\mathcal{I}(\psi) = \{\langle c, d \rangle, \mathfrak{e}\}$, and $\mathcal{V}(\alpha, s) = \{\langle a, b \rangle\}$. Then both φ and ψ hold in $\mathcal{I}$ and α knows φ. But it is plain that α does not know ψ.

---

[6] It is not possible to verify a false proposition.

[7] If one thinks this definition is giving a too strong notion of knowledge, one may think of our attempts as leading only to a notion of awareness, not knowledge per se. A suitable incorporation of an accesibility relation between states would then lead to a version of awareness models. Cf. (Farin, Halpern, Moses, & Vardi, 1995, ch. 9.5.).

The main upshot of this sort of formal representation of knowledge is that the knowledge of 'truths of logic' does not come without effort, so to speak. Let us consider only formulas φ, that have $\mathcal{I}(\varphi) = \{\mathfrak{e}\}$ for every $\mathcal{I}$, call them $\mathfrak{e}$-formulas. Agents have to perform a special procedure to attain knowledge of $\mathfrak{e}$-formulas, namely the procedure $\mathfrak{e}$. However, if an agent comes to know one $\mathfrak{e}$-formula, then he knows them all.[8] This may lead to an interesting alternative solution of the logical omniscience problem.

6 Open problems

We conclude the paper by pointing out the most acute open problems. First, as noted in the section on LSP-validity, our implication connective does not fully represent the notion of P-inclusion: only the trivial direction of the deduction theorem holds, the converse has simple counterexamples. Is it possible to define such an implication connective C, that both directions will be true ($\varphi \supseteq_P \psi$ iff $\models \varphi C \psi$)? We see that some paradoxes of implication hold (e.g., $p \to (q \to q)$), but in the light of the abovementioned problem, this is not a conclusive agrument against the new notion of meaning connetcion embodied in (P).

Second, modelling knowledge with the tools of LSP seems to be promising, but we still lack a proper treatment of formulas like $K_\alpha \varphi$. What is $\mathcal{I}(K_\alpha \varphi)$ for a given φ? How to construct this set from $\mathcal{I}(\varphi)$? Until these problems are solved, LSP can not be viewed as a logic of the meaning connection (P), nor as a proper logic of knowledge as justified true belief.

Yet there is no need to be overly pessimistic. Future work on the epistemic interpretation of LSP could shed light on the procesual understanding of justifications, and thereby offer a novel perspective on the notion of explicit knowledge.

Igor Sedlár
Department of logic and methodology of sience, Comenius University
Šafárikovo námestie 6, 818 01 Bratislava
sedlar@fphil.uniba.sk

---

[8]Similarily for every set of formulas that have the same values under interpretations.

http://www.fphil.uniba.sk

Juraj Podroužek
Institute of philosophy, Slovak Academy of Science
Klemensova 19, 813 64 Bratislava
jurajpodrouzek@gmail.com
http://www.klemens.sav.sk/fiusav/

References

Anderson, A. R., & Belnap, N. D. (1975). *Entailment. the logic of relevance and necessity* (Vol. 1). Princeton: Princeton University Press.

Anderson, A. R., Belnap, N. D., & Dunn, J. M. (1992). *Entailment. The logic of relevance and necessity* (Vol. 2). Princeton: Princeton University Press.

Artemov, S. N. (2001). Explicit provability and constructive semantics. *Bulletin for Symbolic Logic, 7*(1), 1–36.

Artemov, S. N., Kazakov, E., & Shapiro, D. (1999). *On logic of knowledge with justification* (Tech. Rep. No. CFIS 99-12). Ithaca, NY: Cornell University.

Farin, R., Halpern, J. Y., Moses, Y., & Vardi, M. Y. (1995). *Reasoning about knowledge.* Cambridge (Mass.): MIT Press.

Fitting, M. C. (2004). A logic of explicit knowledge. In L. Běhounek & M. Bílková (Eds.), *The logica yearbook 2004* (pp. 1–25). Prague: Filosofia.

Fitting, M. C. (2005). The logic of proofs, semantically. *Annals of Pure and Applied Logic*(135), 1–25.

Fitting, M. C. (2009). Reasoning with justifications. In D. Makinson, J. Malinowski, & H. Wansing (Eds.), *Towards mathematical philosophy* (Vol. 28, pp. 107–123). Berlin: Springer.

Lewis, C. I., & Langford, C. H. (1932). *Symbolic logic.* New York: Century Company.

Mares, E. D. (2004). *Relevant logic. A philosophical interpretation.* Cambridge: Cambridge University Press.

Ajdukiewicz Functions and Basic Inference

Sebastian Sequoiah-Grayson

1 Introduction

That logic concerns interaction just as much as it concerns consequence is an insight has been most eagerly adopted in various logics of multi-agent information flow, such as Public Announcement Logics, and Dynamic Epistemic Logics and so forth ((Benthem, 2009), (Baltag, Moss, & Solecki, 1998), (Baltag & Smetts, 2008), (Ditmarsh, Hoek, & Kooi, 2008)). What we will see, is that paying proper attention to interaction also has a rich payoff for mono–agent, deductive scenarios. The basic idea is this; instead of examining the information flow between different agents in a communicative setting, we examine the information flow between different states of a single agent as the agent reasons deductively. We will see how this approach has its roots in the the pure function application of the Categorical Grammar literature, due to (Ajdukiewicz, 1935), (Lambek, 1958), and (Lambek, 1961).

2 Weakly Commuting Interaction Models

A semantics is *operational* if it allows an explicit representation of semantic operations on individual points in the model (see (Buszkowski, 1986)). In the model below, the points are understood as information states. This though, is still fairly general. Since we are concerned with the information flow in deductive inference, the points are to be understood as information–bearing states of an agent as the agent reasons deductively. This information–bearing is taken to be explicit knowledge. The interplay between the semantics so understood and the resulting syntax is philosophically interesting: The information in a database will be structured in different ways, depending on the use to which the information is to be put. We can think of this structure as a type of database *grammar*. The grammar will set the constraints as to structure of the database. In short, we will specify the grammar of a fragment of the "cognitive langauge" of deductive reasoning.

2.1 A Commutating, Non–Associating Information Frame

Where $A, B, \ldots$ are types of propositional formula $\phi, \psi, \ldots$ such that $\phi\colon A$ is read as *formula ϕ is of type A*, the language of our logic, the *non–associative Lambek Calculus with permutation, bottom, and identity* (**NLP$_{01}$**) is given as follows:

$$A ::= \phi \mid A \mid \mathbf{0} \mid \mathbf{1} \mid A \otimes B \mid A \multimap B \mid A^\perp \qquad (1)$$

Take an information frame $\mathbf{F}\langle S, \sqsubseteq, \bullet \rangle$ with weak (one place) commutation and Cut, along with the two binary connectives $\otimes$, and $\multimap$, the unary connective $^\perp$, and the constants $\mathbf{0}$ and $\mathbf{1}$. S is a set of incomplete, or partial information states $x, y, \ldots$[1] The binary relation $\sqsubseteq$ is a partial order on S of informational development/inclusion. $\bullet$ is the (commutative and non-associative) binary composition operator on information states. $\otimes$ is (commutative and non-associative) internal conjunction (merge/fusion). $\multimap$ is interactive implication, and $^\perp$ is interactive negation on account of its being defined in terms of $\multimap$ and $\mathbf{0}$:

$$A^\perp := A \multimap \mathbf{0}, \qquad (2)$$

$\mathbf{0}$ is bottom, and $\mathbf{1}$ is unit/identity such that:

$$\mathbf{1} \otimes A = A = A \otimes \mathbf{1}. \qquad (3)$$

Hence we have weak commutation around state–combination, and (correspondingly) merge, and weak–weak commutation around identity. Weak–weak commutation around identity is standard enough. Weak commutation (around combination and fusion) allows pairwise one–place swaps: $x \bullet y = y \bullet x$ at least insofar as informational progress is concerned, and $A \otimes B = B \otimes A$ at least insofar as derivability is concerned. That is, for our interactions, commutation is permissible *within* the parameters of the given bracketing. Hence the following are of the same type:

$$A \otimes (B \otimes C) \mid A \otimes (C \otimes B) \mid (C \otimes B) \otimes A \mid (B \otimes C) \otimes A \qquad (4)$$

Whereas:

$$B \otimes (A \otimes C) \qquad (5)$$

---

[1] In doxastic cases, we would allow for inconsistency also. But since knowledge is factive if anything is, we disallow this property here.

differs. Exactly why it is that this should be the case in the context of a significant fragment of mono–agent deductive reasoning is made clear in section 2.4 below. There *is* a difference in other contexts, natural language grammatical ones say, where strict preservation of left–right attachment–detachment is crucial. The natural language semantics cases from categorical grammar are not merely formal relatives. They in fact give us the starting point for the very conceptual framework that we need.

2.2 Categorical Grammars: Pure Sequences, and Pairs

This ordering preservation is due to the fact that natural languages tend to have exact rules for the ordering of words. For example, 'Frederike peddles' is grammatically well–formed, whereas 'Peddles Frederike' is not. In this manner, the intransitive verb 'peddles' is of type $n \rightarrow s$, since when it is applied to the right of a noun (type n), the result is a sentence (type s). By contrast, the adjective 'happy' is of type $n \leftarrow n$. This is due to the fact that when 'happy' is applied to the left of another noun, the result is another (complex) noun (phrase): 'Happy Friederike'. In logics with no commutation (other than possible weak–weak commutation around **1** if **1** is present), the structures are at least pure sequences, and if associativity is rejected also, then they are pairs.

This much is familiar from the categorical grammar literature. What is less familiar is the following useful way of describing the relevant semantic constraints. We can think of natural language (in particular a natural language lexicon) as a *database*. In this case, a grammar may be thought of as a collection of *processing constraints* on the database. These processing constraints are operational in the strictest sense of the term. The tell us which operations are permissible insofar as generating well–formed, or meaningful, strings in the natural language are concerned. In informational terms, they tell us which operations are permissible insofar as generating information is concerned. A detailed look at grammars as processing constraints on a database may be found in (Sequoiah-Grayson, 2009b).

For our purposes here however, all we note is that a natural language lexicon is only one type of database. A consequence of this is that we may think of the processing constraints on *any* database (and not simply natural language lexicons) as *grammars*. This is precisely what is done below with the database relevant to mono–agent deductive reasoning. The rest of this section is dedicated to developing this idea in more detail. This first step is to specify the relevant model.

2.3 The Model

A model $\mathbf{M} := \langle \mathbf{F}, \Vdash \rangle$ is an ordered pair $\mathbf{F} \langle S, \sqsubseteq, \bullet \rangle$ and $\Vdash$ such that $\Vdash$ is an evaluation relation that holds between members of S and formulas constructed out of our binary connectives $\otimes$, and $\multimap$, and constants $\mathbf{0}$ and $\mathbf{1}$. In what follows, we will often write $x, y, \ldots \in \mathbf{F}$ as shorthand for $x, y, \ldots \in S$ where $S \in \mathbf{F}$. Where A is a propositional formula, and $x, y, z \in \mathbf{F}$, $\Vdash$ obeys the heredity or monotonicity condition:[2]

$$\text{For all } A, \text{if } x \Vdash A \text{ and } x \sqsubseteq y, \text{ then } y \Vdash A. \tag{6}$$

And also obeys the following conditions for each of our connectives and constants $\mathbf{0}$ and $\mathbf{1}$:

$x \Vdash A \otimes B$ *iff* for some $y, z, \in \mathbf{F}$ s.t. $y \bullet z \sqsubseteq x$, $y \Vdash A$ and $z \Vdash B$.   (7)

$x \Vdash A \multimap B$ *iff* for all $y, z \in \mathbf{F}$ s.t. $x \bullet y \sqsubseteq z$, if $y \Vdash A$ then $z \Vdash B$.   (8)

$x \Vdash \mathbf{0}$ for no $x \in \mathbf{F}$.   (9)

$x \Vdash A^{\perp}[A \multimap \mathbf{0}]$ *iff* for all $y, z \in \mathbf{F}$ s.t. $x \bullet y \sqsubseteq z$, if $y \Vdash A$ then $z \Vdash \mathbf{0}$.   (10)

$x \Vdash \mathbf{1}$ *iff* $x \in T$, for all $x \in \mathbf{F}$.   (11)

(11) is less straightforward than are (7)–(10). We firstly define the set of *propositions* $\text{Prop}(\mathbf{F})$ on our frame $\mathbf{F}$ as the set of all subsets X of $S \in \mathbf{F}$ such that they are *upwardly closed*: if $x \in X$ and $x \sqsubseteq x'$, then $x' \in X$.

We can now see how it is that T in (11) is a *truth set*. Truth sets come in left and right versions. For any subset T of $\text{Prop}(\mathbf{F})$:

- T is a *left truth set iff* for all $y, z \in \mathbf{F}$, $y \sqsubseteq z$ *iff* for some $x \in T, x \bullet y \sqsubseteq z$.

---

[2] We would drop this condition for certain *doxastic* scenarios where non-monotonicity is a distinctive property.

- T is a *right truth set iff* for all $y, z \in \mathbf{F}$, $y \sqsubseteq z$ *iff* for some $x \in T, y \bullet x \sqsubseteq z$.

Given that we have commutation on our frame, our truth set T is non-directional. In other words, since $x \bullet y = y \bullet x$, the left and right truth sets collapse into a single, non-directional truth set. The converse does not hold. The "for some $x \in T$" constraint in the right hand clause of both truth set conditions allows us to restrict commutation to just these x. This is what allows **1** to commute around propositions in logics that are otherwise non-commutative. Intuitively, the state x carrying the information **1** behaves around other states in the same manner as does **1** around propositions (see (3) in subsection 2.1 above).

With the model conditions laid bare via the operational semantics, we can look at how the various connectives deliver with respect to information processing environments.

2.4 Databases and Information Processing

Our merge and implication connectives interrelate in the following manner:
$$A \otimes B \vdash C \text{ iff } B \vdash A \multimap C. \quad (12)$$

We take $\vdash$ to be a processing–gate in the sense that $X \vdash A$ is read as *processing the information in X generates the information that A*. Exactly what it is that processing amounts to depends on the structure of the database in question, which is in turn fixed by the structural rules at work. Since the merge operation is simply *combination*, and not directional application, we get $A \vdash B \multimap C$ from $A \otimes B \vdash C$ by commuting on $A \otimes B$ so as to get $B \otimes A$. This is one sense in which we depart from Ajdukiewicz's original contribution to categorical grammar, since we do not need to keep track of left–right attachment–detachment. The only structural rule we will admit except for Weak Commutation is Cut. Cut is important for making use of the relevent information interactions at work in basic inference. Association however, is not. To see this, set the following:

$$\phi \Rightarrow \psi \colon A, \sigma \Rightarrow \phi \colon B, \sigma \colon C, \phi \colon D, \psi \colon E. \quad (13)$$

In such a case, we have it that:

$$A \otimes (C \otimes B) \vdash E. \tag{14}$$

(14) is a result of Cutting on D, since:

$$C \otimes B \vdash D, \text{ and } A \otimes D \vdash E. \tag{15}$$

It is in this sense that cut underpins the most fundamental notion of information interaction, or *processing*. However, suppose that we were to *associate* on (14). In this case, we would have it that:

$$(A \otimes C) \otimes B \vdash E. \tag{16}$$

This is not good. The result of combining the information in A with the information in C is nothing such that were it to be applied to the information in B we would get the information in E. In fact, no information results from the combination of information of type A with information of type C. Such an attempt is a "dead process", that cannot be carried out. Allowing multiple types, then given that $\phi \Rightarrow \psi: A$ and $\sigma: C$, it is also the case that $\phi \Rightarrow \psi: C^\perp$. Hence $\phi \Rightarrow \psi: C \multimap \mathbf{0}$ via (2). For any state $x \Vdash C \multimap \mathbf{0}$, we know that it is that case that if we combine this information with any other state y s.t. $y \Vdash A$, then the result will be a state z s.t. $z \Vdash \mathbf{0}$ via (10). However, we know that $\mathbf{0}$ is not supported via any state via (9).

In information processing terms, some information is of type $C^\perp$ iff its combination with information of type C can never generate any information. This is a conceptually parsimonious way of reading the frame condition in (10). Taking interactive implication to be *functional*, along the lines of the Lambek Calculi, then $C^\perp$ is the type of function that can never take information of type C as an input, on account of it never outputting any information on the basis of inputs of this type. This makes perfect sense in our interactive/dynamic setting. In a static setting, negation is ruling out truth. In an interactive/dynamic setting however, negation will rule out particular interactions, or processes.[3]

---

[3] Negation in a dynamic setting as *process exclusion* is a topic unto itself. We can extend the operation into a fully directional process exclusion system by dropping even weak commutation. This will allow us to split our interactive implication $\multimap$ into a double implication pair $\langle \rightarrow, \leftarrow \rangle$, that will in turn allow us to define a split negation pair $\langle \sim, \neg \rangle$ which we may define as $\sim\!A := A \rightarrow \mathbf{0}$ and $\neg A := \mathbf{0} \leftarrow A$ respectively. For an examination of logic–invariant split–negation properties in

Now we can see why it is that a strong, unrestricted (two-place) commutation is just as "de-railing" as associativity. This would allow us to get from $(A \otimes B) \otimes D$ to $(A \otimes D) \otimes B$ – but now just set ϕ: A, $\phi \Rightarrow \psi$: B, ψ: C, $\psi \Rightarrow \sigma$: D, σ: E. Similar results can be obtained for the other structural rules by adjusting the setup in (13) appropriately, see (Sequoiah-Grayson, 2009c). Importantly, strong–commutation and association are not independent.

2.5 Strong–Commutation Recovery and Association Recovery

Commutation with associativity recovers strong–commutation, and commutation with strong–commutation recovers associativity:

- For the first recovery, start with $(A \otimes B) \otimes D$, then associate to get $A \otimes (B \otimes D)$, then commute to get $A \otimes (D \otimes B)$, then associate again to get $(A \otimes D) \otimes B)$. □

- For the second recovery, start with $A \otimes (B \otimes D)$, and strongly commute to get $B \otimes (A \otimes D)$, then commute to get $(A \otimes D) \otimes B$, then strongly–commute to get $(A \otimes B) \otimes D$. □

In otherwords, we have generative as well as independent reasons to reject associativity and strong–commutation.

With this much done, we are in a position to specify the processing-constraints, or grammar of a fragment of mono–agent deductive reasoning.

2.6 The Grammar of Deductive Reasoning

The *grammar* of deductive reasoning, a cognitive–language if you will, has obvious fragments with useful properties captured by a commutating, non–associating logic. The logic corresponds to **NLP**: the *non-associative Lambek Calculus with permutation*, where permutation is understood in the pair–wise sense such that is amounts to:

---

terms of process exclusion, as well as an examination of its philosophical status, see (Sequoiah-Grayson, 2009a). For a working through of a series of examples of process exclusion, both directional and non–directional, as well as an examination of the connections with the related notion of negation as test–failure in *dynamic predicate logic*, see (Sequoiah-Grayson, 2009b).

$$x \bullet y \sqsubseteq z = y \bullet x \sqsubseteq z. \tag{17}$$

Specifying explicitly the pairwise nature of the permuting operation is not redundant. This is because it is commonplace in the literature to use 'permutation' to denote the strong–commutation that follows from commutation, or pairwise permutation, and association. This is a simple function of the fact that commutating, non-associating logics have are rare, so the resulting stronger permuting operation, allowing permutations through bracketing, or *structure*, has been the default.

The structure of the data–base on the left–hand–side of the processing gate ⊢, that is X in $X \vdash A$ that is specified by the grammar is that of *mobiles*. Mobiles are simply non-associating but bracket–sensitive–commutating structures. Since we can get strong–commutation from commutation if we also had association, the addition of association would collapse the structure of our data–base into multisets (since we do not have contraction). But multisets have no structure at all, they have merely a taxonomy. Mobiles have some structure, but less than do pure sequences (where even pair–wise commutation is prohibited). If we had neither commutation nor association, we would have *static pairs*, with a fixed left component and a fixed right component. Mobile may be thought of as mobile pairs, which is what they really are after all; pairs whose left and right components may switch places.

Since we have bottom, **0**, and unit, **1**, we will denote $\mathbf{NLP} + \mathbf{0} + \mathbf{1}$ as $\mathbf{NLP_{01}}$. With the behaviour of the logic laid out, it is now time to put it to good use.

3 Mono–Agent Dynamic Reasoning

It is important to appreciate that what is happening here is *not* simply syntactic (although it can be read off the syntax, which is part of the appeal). Merge (along with our implication) is being interpreted as a relation between *information states*. It is in this sense that the interaction structures (such as that on the left hand side of (14): $A \otimes (C \otimes B)$) are robustly semantic. Two interaction structures with the same form will not necessarily be equivalent, since they may be underpinned by different information states. Extracting the step–wise information state combinations across S *corresponding* to the relevant interaction structures is a straightforward mechanical task involving

Ajdukiewicz Functions and Basic Inference

nothing more than successive applications of the conditions outlined by (7) above. This is most easily seen via some examples.

3.1 Interaction Structures and Processing Structures

With respect to (14): $A \otimes (C \otimes B) \vdash E$, we have the following corresponding step–wise information state combination:

$x \Vdash A \otimes (C \otimes B)$ iff for some $w, y, z \in \mathbf{F}$ s.t.
$$w \bullet (y \bullet z) \sqsubseteq x, \ w \Vdash A, y \Vdash C, \text{ and } z \Vdash B. \quad (18)$$

The information states $x, y, \ldots \in S$ may be naturally interpreted as states of α as α reasons deductively. In this case, the information state combination $w \bullet (y \bullet z)[\sqsubseteq x]$ specifies the step–wise reasoning procedure that α must engage in in order to be truthfully said to know (on the basis of the premises at least) the result of the merged propositions, namely ψ. Since, $y \Vdash C$, and $z \Vdash B$ via (18), and $\sigma : C$ and $\sigma \Rightarrow \phi : B$ via (13), $y \bullet z \sqsubseteq v$, where $v \Vdash D$, and $\phi : D$ via (13). Since $w \Vdash A$ via (18) and $\phi \Rightarrow \psi : A$ via (13), $w \bullet v \sqsubseteq x$, where $x \Vdash E$ and $\psi : E$ via (13). □

Via (12), we can transform interaction structures into iterated conditional information *processing structures*. Still taking (14) as our case, via three applications of (12) and one application of (3), we generate:

$$1 \vdash B \multimap (C \multimap (A \multimap E)). \quad (19)$$

From $A \otimes (C \otimes B) \vdash E$ we get $C \otimes B \vdash A \multimap E$ via the first application of (12). From $C \otimes B \vdash A \multimap E$ we get $B \vdash C \multimap (A \multimap E)$ via the second application of (12). From $B \vdash C \multimap (A \multimap E)$ we get $B \otimes 1 \vdash C \multimap (A \multimap E)$ via (3). From $B \otimes 1 \vdash C \multimap (A \multimap E)$ we apply our third and final instance of (12) in order to get $1 \vdash B \multimap (C \multimap (A \multimap E))$. □

We can understand the processing structure on the right hand side of (19) as a *typed function*. It is the type of function that takes inputs of type B, and returns another function as the output. The function that it returns as an output is the type of function that takes inputs of type C, and returns yet another function as an output. *This* function is the type of function that takes inputs of type A and returns an output of type E (which in our case is ψ, since $\psi : E$ via (13)).

Similarly to interaction structures, the processing structures/function types are individuated by information states. With respect to (19), and via (8), we have the following:

$$x \Vdash B \multimap (C \multimap (A \multimap E)) \text{ iff}$$
$$\text{for all } s, t, v, w, y, z \in \mathbf{F} \text{ s.t. } ((x \bullet y) \bullet v) \bullet t \sqsubseteq s, \quad (20)$$
$$\text{if } z \Vdash C \multimap (A \multimap E), \text{ and } y \Vdash B, \text{ and } w \Vdash A \multimap E,$$
$$\text{and } v \Vdash C, \text{ and } t \Vdash A, \text{ then } s \Vdash E.$$

Since $x \bullet y \sqsubseteq z, z \bullet v \sqsubseteq w$, and $w \bullet t \sqsubseteq s$. □

We can now turn our attention to examining the conceptual relationship between interaction structures and processing structures on the one hand, and mono-agent reasoning on the other. The following section explains how it is that we may sensibly interpret interaction structures and processing structures as executions and instructions respectively.

3.2 Instructions and Executions

How might we think of the interaction structures such as that in (18) and their corresponding processing structures such as that in (20) with respect to our wider concern with interactive mono-agent reasoning? We can think of processing structures as *instructions*, and of their corresponding interaction structures as the result of carrying out or executing the corresponding instruction, i.e., as *executions*.

Take the instruction in (20). If α is in state x, then α is on her way to knowing explicitly *that* ψ, but she is not there yet. α's being in state x means that α knows explicitly what is required in order that she come to know explicitly *that* ψ. Her first step is to establish $B(\sigma \Rightarrow \phi)$, which involves her being in state y. Her second step is to combine this state y with her previous state x. In other words, we read the iterated state–combination sequence corresponding to process structures such as $((x \bullet y) \bullet v) \bullet t[\sqsubseteq s]$ in (20) from the inside–out.

This first interaction or merge will result in two things. α will know explicitly *that* B, and also be in a new state $z \Vdash C \multimap (A \multimap E)$ which follows from this first interaction. This corresponds to the first leftwards–transfer across the processing–gate. That is, α has moved from $1 \vdash B \multimap (C \multimap (A \multimap E))$ to $B \otimes 1 \vdash (C \multimap (A \multimap E))$

with the establishing of B allowing α to get rid of $\mathbf{1}$ and arrive at $B \vdash (C \multimap (A \multimap E))$.

How might we interpret $\mathbf{1}$? That is, how might we make sense of the initial state x of α s.t. $x \Vdash \mathbf{1}$? We do so in exactly the sense stipulated above; α being in state x which carries the information that $\mathbf{1}$, is simply that state in which α knows what steps must be taken in order to establish E, and hence be in state s.

The next step for α is for her to establish $C(\sigma)$, which entails that she be in state $v \Vdash C$. By combining v with z, α will then be in state $w \Vdash A \multimap E$. This corresponds to the second leftwards–transfer across the processing–gate, such that α has moved from $B \vdash (C \multimap (A \multimap E))$ to $C \otimes B \vdash A \multimap E$. The final steps for α are that she establish $A(\phi \Rightarrow \psi)$, which entails her being in state $t \Vdash A$. Then α must combine t with her previous state w. This will entail α being in state $s \Vdash E$, where $\psi : E$. This corresponds to the third leftwards move across the processing–gate, such that α has moved from $C \otimes B \vdash A \multimap E$, to $A \otimes (C \otimes B) \vdash E$.

What has occurred is this: By following the instructions laid out in the processing structure, α has extracted the very interaction structure who's "activation" will cause her to know explicitly *that* ψ. This fact has a straightforward interpretation in terms of the data–base structure, or grammar, of the "cognitive langauge" of deductive reasoning. The interpretation of the end-results of applications of the residuation conditions in (12) as instructions and executions is a powerful one. The instructions tell us how to construct a well-formed sentence in the cognitive language of deductive reasoning in precisely the same manner as the types in categorical grammar instruct us on constructing well-formed terms or sentences in natural language. In natural language, if the result of one of these constructions is uttered by an agent, then information is transmitted to other agents. Something meaningful, or informationally well-behaved, will have occurred. Even if not uttered of course, something meaningful, or informationally well-behaved will have occurred.

The deductive reasoning case is the same. When the results of carrying out the instructions are executed, something informationally well-behaved has occurred. In this case, the occurrence is that the agent has accessed the information in the conclusion, or simply put; the agent is in a state such that the agent knows explicitly the information that the conclusion encodes. That deductive reasoning procedures should have their behaviour accurately described by formal tools originally developed to specify that mathematical behaviour of natural language semantics is, ultimately, not as surprising as it might first appear. After all, when we are engaged in explicit acts of deductive reasoning, it is rather like we are talking to ourselves.

4 Conclusion

By paying proper attention to consequence *and* interaction in logic, we have a system of resource–management and processing on the database of mono–agent reasoning dynamics.[4]

Sebastian Sequoiah-Grayson
Postdoctoral Research Fellow, Formal Epistemology Project
Centre for Logic and Analytical Philosophy
University of Leuven
Kardinaal Mercierplein 2, 3000 Leuven, Belgium
seb.sequoiahgrayson@hiw.kuleuven.be
logic.tsd.net.au

References

Ajdukiewicz, K. (1935). Die Syntaktische Konnexitat. *Studia Philosophica*, 1–27.

Baltag, A., Moss, L., & Solecki, S. (1998). The logic of public announcements, common knowledge and private suspicions. In *Proceedings of tark '98* (pp. 43–56). Evanson, IL: Morgan Kaufmann Publishers.

Baltag, A., & Smetts, S. (2008). The logic of conditional doxastic actions. In R. van Rooij & K. Apt (Eds.), *Texts in logic and games*. Amsterdam: Amsterdam University Press. (Special issue on New Perspectives on Games and Interaction. Forthcoming in)

---

[4]I am greatly indebted to Johan van Benthem for first suggesting to me that a deeper examination of the Categorical Grammar literature may make a positive contribution to the analysis of mono–agent dynamics. As always, his advice was good. I am also indebted to Greg Restall for answering endless questions about various commutation types. I would also like to thank Jake Chandler and Marie Duzi for helpful comments and suggestions. Any remaining mistakes remain my own.

Benthem, J. van. (2009). *Logical dynamics of information and interaction*. (Manuscript)

Blackburn, P., Rijke, M. de, & Venema, Y. (2002). *Modal logic*. Cambridge: Cambridge University Press.

Buszkowski, W. (1986). Completeness results for Lambek syntactic calculus. *Zeitschrift für Mathematische Logik und Grundlagen der Mathematik*, *32*, 13–28.

Ditmarsh, H. van, Hoek, W. van der, & Kooi, B. (2008). *Dynamic epistemic logic*. Berlin–New York: Springer.

Lambek, J. (1958). The mathematics of sentence structure. *American Mathematical Monthly*, *65*, 154–170.

Lambek, J. (1961). On the calculus of syntactic types. In R. Jakobson (Ed.), *Structure of language and its mathematical aspects* (pp. 166–178). Providence, RI: American Mathematical Society.

Moortgat, M. (1995). *Residuation in mixed Lambek systems* (Tech. Rep. No. Research Transcript no. 10). Utrect: Research Institute for Language and Speech (OTS).

Sequoiah-Grayson, S. (2009a). Dynamic negation and negative information. *Review of Symbolic Logic*, *2*(1), 233–248.

Sequoiah-Grayson, S. (2009b). *Lambek calculi with **0** and test–failure in DPL*. (Forthcoming in *Linguistic Analysis*.)

Sequoiah-Grayson, S. (2009c). A positive information logic for inferential information. *Synthese*, *167*(2), 409–431.

Ontological Discriminations

Hartley B. Slater

1

Russell held that 'a exists', where 'a' is a logically proper name, was necessarily true. By contrast his account of 'The K exists' allowed this to be contingent, since, on his Theory of Descriptions, it did not assert the existence of an individual, but merely the instantiation of some uniquely identifying properties. The present paper refines Russell's distinction in several ways, first by providing what Russell merely gestured at, namely explicit, formally defined logically proper names. But following from this it is seen that Russell's intention with regard to 'The K exists' is better expressed 'A unique K exists', leaving the former to be assimilated into the non-contingent category, through interpreting its subject phrase 'The K' non-attributively. The paper closes with an exhibition of similar discriminations that are available with higher-order subjects, such as properties, numbers, and facts; and most particularly with propositions.

We must first have clear what referential terms are. This was obscured in Russell's Theory of Descriptions, and remains unclarified to this day because of the popularity of that theory, and the associated neglect of Hilbert's Epsilon Calculus. In Russell's Theory there are, it will be remembered, three clauses with 'The King of France is bald'. These are 'there is a king of France', 'there is only one king of France' and 'he is bald'. Russell used an iota term to symbolise the definite description, but it is not an individual symbol: it is an 'incomplete' term, as he explained it, since 'The King of France is bald' is taken to have the complex analysis, '$(\exists x)(Kx.(y)(Ky \supset y = x).Bx)$', and so it does not have the elementary form 'Bx'. Russell hypothesised that, in addition to the linguistic expressions gaining formalisations by means of his iota terms, there was another, quite distinct class of expressions, which he called 'logically proper names'. Logically proper names would, amongst other things, take the place of the variable in such forms as 'Bx'. Russell suggested that demonstratives might be in this class, but he could give no further formal expression to them. The epsilon theory of descriptions that settles the question was discussed in the first edition of Hughes and Cresswell's classic introductory text on Modal Logic, and originated with Routley, Meyer and Goddard, who, in their work on intensional contexts, made an explicit identification of definite descriptions with epsilon terms: The King of France $= \varepsilon x(Kx.(y)(Ky \supset y = x))$ (Goddard & Routley, 1973, p. 558) and (Hughes & Cresswell, 1968, p. 203).

Which theorems in the epsilon calculus are behind this kind of identification? The standard epsilon calculus contains the axiom '$(\exists x)Fx \supset F\varepsilon xFx$' (Leisenring, 1969), (Meyer Viol, 1995), from which one can naturally obtain the equivalence between the two sides. There is then one theorem in particular which demonstrates strikingly the relation between Russell's attributive, and Donnellan's 'purely referential' understanding of referential terms. For

$$(\exists x)(Kx.(y)(Ky \supset y = x).Bx), \qquad (1)$$

is logically equivalent

$$(\exists x)(Kx.(y)(Ky \supset y = x)).Ba, \qquad (2)$$

where $a = \varepsilon x(Kx.(y)(Ky \supset y = x))$, this being the epsilon term arising from the first conjunct in (2). The first expression, as we have seen, encapsulates Russell's Theory of Descriptions, in connection with 'The K is B'; it involves the explicit assertion of the first two clauses, to do with the existence and uniqueness of an K. Since Donnellan, however (Donnellan, 1966), we have realized that there are no preconditions on the introduction of 'the K' as an individual term. So 'The K is B', with 'The K' an individual term, may always be given a truth value, even if, sometimes, that truth value is merely an arbitrarily chosen one. For 'Ba' properly formalises 'The K is B', since the cross reference in (2) means that it reads 'There is a single K. It is B', and the descriptive replacement for the E-type pronoun 'it', there, is 'The K'. If the description in 'a' is non-attributive, i.e., if the first two clauses of Russell's account are not both true, and there is no such thing as the King of France, then the referent of 'the K' is simply up to the speaker to nominate — perhaps as a private, or secret name for something.

How can something be the one and only K 'if there is no such thing'? That is where a second, and even more important theorem in the epsilon calculus is required:

$$(Ka.(y)(Ky \supset y = a)) \supset [a = \varepsilon x(Kx.(y)(Ky \supset y = x))].$$

For the singular thing is that this entailment cannot be reversed, so there is a difference between the left hand side and the right hand side, i.e. between something being alone king of France, and that thing being the one and only king of France. The difference is not available in Russell's logic, since only possession of the property can be formalised there. In fact Russell confused the two forms, since possession of an identifying property he formalised using the identity sign, viz '$a = \iota x K x$', making it appear that some, maybe even all identities are contingent. It is this idea that has been incorporated into Free Logics, and so we start to see the inappropriateness of them. For all proper identities are necessary, and it is merely associated identifying properties that are contingent. That means that in all possible worlds there is the same domain of discourse, although the individuals in that domain may change their properties, and even their individuating properties, from one world to the next. One specific form of this distinction is of particular moment in what follows: the distinction between possession of the same name, and being the same individual. We shall find that traditional modal semantics has confused the two, and that only their separation can give a clearer picture of identity in modal contexts.

2

What are the consequences of the above for modal, and general intensional logics? Clearly, if the same individual is to be involved, epsilon terms must be rigid across all worlds. Richard Routley presented several rigid intensional semantics, some objectival and some substitutional (Routley, 1977, pp. 185–186). One of these semantics, for instance, simply took the first epsilon axiom to hold in any interpretation, and made the value of any epsilon term itself. Using such a rigid semantics, Routley, Meyer and Goddard obtained what has been called 'Routley's Formula', i.e. '$L(\exists x)Fx \supset (\exists x)LFx$', by means of the following transformations. Routley's Formula holds for any propositional operator and any predicate, but they illustrated it in the case of necessity, and with 'Fx' as 'x numbers the planets'. Then, with '$\varepsilon x F x$' as 'the number of the planets', they said that from '$L(\exists x)Fx$', we can get '$LF\varepsilon x F x$', by the epsilon definition of the existential quantifier, and so '$(\exists x)LFx$', by existential generalisation over the rigid term (Routley, Meyer, & Goddard, 1974, p. 308); (Hughes & Cresswell, 1968, pp. 197 and 204). (For further discussion of Routley's Formula see, for instance, (Slater, 1992)).

We therefore see not only that $L(\exists x)(x = a)$, because it is provable that something is a, but also that $(\exists x)L(x = a)$, i.e., that the same thing is a in all possible worlds. In Kripkean semantics it seems we can discriminate between '$L(\exists x)(x = a)$' and '$(\exists x)L(x = a)$' on the grounds that the former merely says that in every possible world there is something named 'a', while the latter says that the same thing is named 'a' in every possible world. But this way of looking at the formulae, we now see, involves a use-mention confusion. Certainly one can discriminate between 'in every possible world something is named 'a'', and 'the same thing is named 'a' in every possible world', but these are meta-linguistic, relational remarks, which ought to be symbolised '$(y)(\exists x)Nx`a`y$', and '$(\exists x)(y)Nx`a`y$', where '$Nx`a`y$' says that x is named 'a' in world y.

One further consequence is the main one signalled at the start, that individuals, properly so-called, have eternal existence, and so must be separated from any entities that merely have 'existence' in this world, or some other, cf. (Stalnaker, 2003, pp. 120–126). What, indeed, in relation to individuals, has 'existence' just in this world, or just in some other — making them 'physical objects', and 'fictions', respectively — are *identifying properties*. Aristotelian Realism therefore holds for such properties, whereas Platonic Realism holds for the associated individuals. It is more usual, perhaps, to think of Platonic Realism and Aristotelian Realism as being rivals, in opposition to one another, because they are seemingly differing accounts of the same thing. But here we come to see that they are merely complementary, through being concerned with different things. The point is illustrated most clearly in the second theorem above. For the left-hand-side entails that certain identifying properties are instantiated, while the crucially different right-hand-side instead entails that a certain object exists — eternally.

The point is central to understanding how eternally real objects are accessed — a seemingly perennial difficulty with Platonic entities. Paradigmatically the situation is represented in the epsilon variant to Russell's analysis of 'The king of France is bald'. For the epsilon variant to Russell's '$(\exists x)(Kx.(y)(Ky \supset y = x).Bx)$', is

$$`(\exists x)(Kx.(y)(Ky \supset y = x)).B\varepsilon x(Kx.(y)(Ky \supset y = x))\text{'},$$

as we saw. But through presenting Russell's formula as this conjunction we enable a separation to be made between a true or false assertion about this world, namely the first conjunct delimiting existence and uniqueness conditions, and a further assertion, in the second conjunct, which is made about its subject independently of whether the first conjunct is true or false, and so about something that exists eternally. It is the first conjunct that tells us whether certain individuating properties 'exist' in the Aristotelian sense, and so properly that conjunct, should be read 'A sole king of France exists', not 'The sole king of France exists'. 'The sole king of France exists' should be formalized instead,

$$`(\exists y)(y = \varepsilon x(Kx.(z)(Kz \supset z = x)))\text{'},$$

which is a straight theorem of the epsilon calculus, and therefore necessary. Access to the object $\varepsilon x(Kx.(z)(Kz \supset z = x))$ is provided entirely by means of the linguistic act of supposing there is a sole king of France, and through its then being invariably possible to cross-refer to the same individual from within further assertions. Eternal objects, in this way, are simply subjects of discourse.

3

The above presents the required details in the case of first-order subjects; by analogy with it, parallel things hold with higher-order subjects. So we can now also show how the commonly supposed abstraction of higher-order subjects, like properties, facts, and numbers has to be re-thought. For if someone is happy, then Happiness in a certain sense exists in the physical world, just like any other physical thing — and likewise with the fact that someone is happy, and with the number of people that are happy — for instance, the 10 people that are happy. But if no-one is happy then Happiness can, in a clear sense, only be dreamt about, and both the fact that someone is happy and the ten people that are happy are fictions. So whether properties, facts, and numbers of things are 'abstract' or 'real' objects might also be said to be contingent. Following (Cocchiarella, 1986) I shall formalise the property of being happy as $\lambda y H y$, the property of being 10 in number as $\lambda P(10y)Py$, the proposition that someone is happy as $\lambda(\exists z)Hz$, and so the fact that someone is happy as $\varepsilon y(Ty.y = \lambda(\exists z)Hz)$.

Then it is contingent whether $(\exists x)Hx$, but not whether $(\exists x)(x = \lambda y Hy)$, and it is contingent whether $(10x)Hx$, but not contingent whether $(\exists x)(x = \lambda P(10y)Py)$. Also it is not contingent whether

$$(\exists x)(x = \varepsilon y(Ty.y = \lambda(\exists z)Hz)),$$

or contingent whether

$$(\exists x)(x = \lambda(\exists y)Hy),$$

but it is contingent whether

$$(\exists y)(Ty.y = \lambda(\exists x)Hx) \quad (\equiv T\lambda(\exists x)Hx, \equiv (\exists x)Hx).$$

Of course, if there is a property, a natural number, or a fact, that does not arise in the actual world in these ways, then it would be counted as unreal in the Aristotelian sense, and so could be said to be 'abstract' or 'fictional'. But always, with respect to the right kind of object, associated with Platonic Realism clearly, there is no contingency. And with each such object access to it is merely through allusion to it, in the appropriate language. Thus the fact that someone is happy, i.e. $\varepsilon y(Ty.y = \lambda(\exists z)Hz)(= h)$, is alluded to when it is claimed that it is true that someone is happy i.e. that $(\exists y)(Ty.y = \lambda(\exists x)Hx)$ $(\equiv (Th.h = \lambda(\exists x)Hx))$. But the fact exists to be talked about even if it is not true that someone is happy.

4

More dramatically, with propositional referential terms in hand we can now not only symbolise many forms of speech which have been overlooked since Quine and before, but also validate many forms of inference that have been hidden from view during the same period. (In fact we can also resolve all paradoxes in the Liar family, as is shown in (Slater, 2010).)

For one quite new form of expression comes into logical theory as a result of having referring terms to propositions, though: propositional identities of the form '$x = \lambda p$'. There are many such with names in place of 'x' — for instance those explicatiing the substance of Pythagoras' Theorem, Goldbach's Conjecture, The Peter Principle, and Murphy's Law. Quite commonly, however, the present generation of logical theorists have tried to insert a variety of set-theoretic expressions in place of 'x'. But propositions are not sets of any kind, since propositions are the referents of nominalised sentences. A nominalising functor is needed, therefore, and the one involving the extension of the Lambda Calculus that Cocchiarella provided recognises most clearly the true continuity with concepts developed in the past.

Using this new language it is then easy to symbolise, and validate many intensional inferences in natural language which previously have been left unformalised. Thus from 'My proposal is that we go to Benidorm for our holiday', (i.e. '$m = \lambda p$') with 'You accept my proposal', (i.e. 'Aym'), we can get 'You accept that we go to Benidorm for our holiday', (i.e. '$Ay\lambda p$'). It is significant that this inference cannot be validated using an operator expression for 'you accept that', since then nothing of the form 'Aym' or '$Ay\lambda p$' would be available in which substitution of the original identity could take place. An operator expression, such as 'Oyp', would not discriminate a place for a nominal expression referring to a proposition, like 'm' or 'λp', since it would fuse the associated nominaliser 'that' with the verb.

But there are also quantificational discriminations between operator forms and referential forms that are highly significant. For we can now symbolise, first of all, *entailments*: '$A \to B$' is not 'A entails B' (for any '$\to$'), since 'entails' is a verb, and so needs nominal expressions on either side of it; properly it is 'that A entails that B', i.e. '$\lambda AE\lambda B$' ($\equiv L(A \supset B)$). That is not just a minor grammatical matter, however, since without the nominalising facility provided by 'that'-clauses, and similar referential phrases, there are major difficulties when one turns to modal logic. Thus, in that area, many might follow Prior (c.f. (Leitgeb, 2009)), who did not use such referential constructions, and say that it was provable that

$$(\exists p)([L(p \supset B) \vee L(p \supset \neg B)].p).$$

For instances of its matrix (call it 'Cp') seemingly follow from B, and also from $\neg B$. But if there are propositional epsilon terms in this style, i.e., if $(\exists p)Cp \equiv C\varepsilon pCp$, then we could prove that

$$[L(\varepsilon pCp \supset B) \vee L(\varepsilon pCp \supset \neg B)].\varepsilon pCp,$$

and so, by Necessitation, that $L\varepsilon pCp$, and by rule K, that

$$(L\varepsilon pCp \supset LB) \vee (L\varepsilon pCp \supset L\neg B).$$

Hence we would get that $LB \vee \neg L \neg B$ (for any 'B'), and so modal collapse. But if one uses referential propositional quantification, no such conclusion can be drawn. For then 'LB' can be represented '$N\lambda B$', with 'N' the predicate of 'that'-clauses and similar nominals, 'is necessary'. Likewise 'Tr' says that r is true, with $T\lambda A \equiv A$, and rule K becomes: if rEs then $Nr \supset Ns$. Certainly then we can say that there is a true proposition which either entails that B or entails that $\neg B$, i.e.

$$(\exists r)([rE\lambda B \vee rE\lambda \neg B].Tr).$$

But if the matrix there is 'Dr' then, in the referential epsilon equivalent, we only get that

$$[\varepsilon r Dr E\lambda B \vee \varepsilon r Dr E\lambda \neg B].T\varepsilon r Dr,$$

and so that $LT\varepsilon r Dr$, and that

$$(N\varepsilon r Dr \supset N\lambda B) \vee (N\varepsilon r Dr \supset N\lambda \neg B).$$

So $N\lambda B \vee N\lambda \neg B$ is not obtainable.

If $\varepsilon r Dr$ was necessary ($N\varepsilon r Dr$) we could again draw the conclusion that there was modal collapse, but all that is necessary is that $\varepsilon r Dr$ is true ($LT\varepsilon r Dr$, i.e. $N\lambda T\varepsilon r Dr$), which allows $\varepsilon r Dr$ to be contingent. In understanding this difference it is important to remember that the referential epsilon term is not itself an expressed proposition, but merely a pro-form referring to the proposition alluded to in the existentially quantified form before.

Hartley B. Slater
Philosophy M207
University of Western Australia
35 Stirling Highway, Crawley
W. A. 6009, Australia
hartley.slater@uwa.edu.au

References

Cocchiarella, N. (1986). *Logical investigations of predication theory and the problem of universals.* Naples: Bibliopolis.

Donnellan, K. (1966). Reference and definite descriptions. *Philosophical Review*, 75, 281–304.

Goddard, L., & Routley, R. (1973). *The logic of significance and context*. Aberdeen: Scottish Academic Press.

Hughes, G., & Cresswell, J., M.˙ (1968). *An introduction to modal logic*. London: Methuen.

Leisenring, A. (1969). *Mathematical logic and Hilbert's epsilon symbol*. London: Macdonald.

Leitgeb, H. (2009). Formal and informal provability. In O. Bueno & O. Linnebo (Eds.), *New waves in philosophy of mathematics* (pp. 263–299). New York: Palgrave Macmillan.

Meyer Viol, W. (1995). *Instantial logic*. Amsterdam: ILLC.

Routley, R. (1977). Choice and descriptions in enriche intensional languages 1, 2, 3. In E. Morscher, J. Czermak, & P. Weingartner (Eds.), *Problems in logic and ontology*. Graz: Akademische Druck- und Velagsanstalt.

Routley, R., Meyer, R., & Goddard, L. (1974). Choice and descriptions in enriched intensional languages 1. *Journal of Philosophical Logic*, *3*, 291–316.

Slater, B. (1992). Routley's formulation of transparency. *History and Philosophy of Logic*, *13*, 215–224.

Slater, B. (2010). What priest (amongst many others) has been missing. *Ratio*, *XXIII*(2).

Stalnaker, R. (2003). *Ways a world might be: Metaphysical and anti-metaphysical essays*. Oxford: Clarendon.

Philosophical Aspects of Display Logic

Shawn Standefer*

Logical inferentialism is the view that the rules governing a connective in a proof-theoretic framework specify or give the meaning of that connective. In natural deduction systems there are two sorts of rules, introduction and elimination rules, which correspond to left and right introduction rules in consecution calculuses, respectively. It is usually expected that there be an appropriate "fit" between the rules for a connective, which is known as the condition of *harmony*. In Dummett's original formulation, harmony says that neither the grounds for introducing a connective nor the consequences of eliminating that connective should outstrip the other. There have been many criticisms of Dummett's formulation and many new proposals. Discussions of inferentialism and harmony are often conducted against the background of natural deduction systems.

We begin with some background on display logic (§ 1). We will then argue that consecution calculuses, rather than natural deduction systems, are the proper settings for philosophical reflection on the meaning of connectives (§ 2). These arguments support using the display logic generalization, in particular (§ 3). We will conclude by presenting a novel characterization of harmony with some applications (§ 4).

1 Background on display logic

We begin with some brief background on display logic. Display logic, which was developed by Belnap, is a generalization of the consecution calculus.[1] Consecutions in standard consecution calculuses, hereafter called "Gentzen-style," are sequences or multisets of formulas on either side of a turnstile sign. Rules in Gentzen-style systems may con-

---

*I have benefitted greatly from discussions of these issues with many people, including Ole Hjortland, Kohei Kishida, Tyke Nunez, Milos Vuletic, Greg Restall, Jarda Peregrin, Stephen Read and Heinrich Wansing. I especially thank Nuel Belnap and Bob Brandom for their philosophical and financial support.

[1] The original presentation is reprinted with some amendments as (Belnap, 1992).

tain formulas on either side of the turnstile that are not individually displayed, indicated by 'Γ' and 'Δ'.

$$\frac{\Gamma, A, B \vdash \Delta}{\Gamma, A\&B \vdash \Delta} \text{ (\& $\vdash$)} \qquad \frac{\Gamma \vdash A, \Delta \quad \Gamma \vdash B, \Delta}{\Gamma \vdash A\&B, \Delta} \text{ ($\vdash$ \&)}$$

Often, Gentzen-style systems have separate structural rules, whose general form does not display any logical connectives, as in the following examples.

$$\frac{\Gamma \vdash \Delta}{\Gamma, A \vdash \Delta} \text{ ($K \vdash$)} \qquad \frac{\Gamma \vdash A, A, \Delta}{\Gamma \vdash A, \Delta} \text{ ($\vdash W$)}$$

Display logic uses two sorts of elements, *formulas* and *structures*. Formulas are defined as one would expect. Structures are built out of formulas and structural connectives, such as the binary connective ∘, the unary negative ∗, the unary modal •, and the null-ary I. Consecutions in display logic consist of a turnstile sign with a single structure on each side.[2]

All structures have a *polarity*, either antecedent or consequent, depending on their positions. Structures on the left of the turnstile have antecedent polarity and structures on the right of the turnstile have consequent polarity. The ∗ connective switches the polarity of all structures in its scope. Every structural connective is subject to *display equivalences*, which are rules that allow a structure to be displayed as the sole element on the side of the turnstile corresponding to its polarity. The following consecutions, for example, are display equivalent.

$$X \vdash Y \circ Z \Longleftrightarrow X \circ Y^* \vdash Z \Longleftrightarrow X \vdash Z \circ Y.$$

There are separate *structural rules* for structural connectives. Depending on which logic is desired, different sets of postulates can be chosen for the structural connectives, which are then indexed according to the rules they obey. The following is an example of a structural rule.

$$\frac{X \vdash Z}{Y \circ X \vdash Z} \text{ ($K \vdash$)} \qquad \frac{(X \circ Y) \circ Y \vdash Z}{X \circ Y \vdash Z} \text{ (W)}$$

Finally, the *connective rules* introduce logical connectives. These have the feature that the *principal constituent* of the inference, the formula

---

[2] We follow the convention of using variables from the end of the alphabet for structures and variables from the beginning of the alphabet for formulas.

Philosophical Aspects of Display Logic

in which the connective in question is the main operator, is displayed as the sole element on the appropriate side of the turnstile. The following pair of rules are the connective rules for &.

$$\frac{A \circ B \vdash X}{A \& B \vdash X} \; (\& \vdash) \qquad \frac{X \vdash A \quad Y \vdash B}{X \circ Y \vdash A \& B} \; (\vdash \&)$$

Our account of harmony, which we will present in §4, is based on two general theorems of display logic, the *identity theorem* and the *elimination theorem*.[3] The identity theorem says that

$$A \vdash A$$

is derivable for all formulas A. In display logic the only axioms postulated are those for atomic p, $p \vdash p$. This theorem would be trivial if $A \vdash A$ were an axiom for arbitrary A. Rather, the theorem is proved by induction on the complexity of formulas. The main connective in the formula on the left of the turnstile must be introduced by the left introduction rule, and similarly for the main connective to the right of the turnstile. The proof of the identity theorem relies only on the formula connective rules and display equivalences; structural rules are not used.

The elimination theorem says that cut

$$\frac{X \vdash M \quad M \vdash Y}{X \vdash Y}$$

is admissible, for structures X and Y and formulas M. The display logic elimination theorem shows that nothing new is provable upon the addition of cut to an otherwise cut-free system. The proof uses the notion of an analysis of rules.[4] In (Belnap, 1992), the proof relies on the fact that in display logic, the rules satisfy eight conditions on analyses, most of which are verified by checking that the rules have the proper form. Of particular importance for us is condition eight, which says that when the cut formula is principal in both premises, the conclusion of cut is obtainable from its premises using cut restricted to proper subformulas, and it need not use any structural rules. The proof

---

[3] In this paper we follow (Belnap, 1992) in using the name "elimination theorem" rather than "cut elimination theorem." This has a precedent in (Curry, 1977).

[4] The idea for analyses comes from (Curry, 1977).

of the elimination theorem itself appeals only to the eight conditions, so any logic that can be given a display formulation will automatically satisfy the elimination theorem.

That is sufficient background on display logic. We will proceed to the argument that consecution calculuses are preferable as backgrounds for inferentialism.

2 Advantages of consecutions

Many discussions of logical inferentialism take place in the context of natural deduction systems. We will argue that consecution calculuses are preferable as a backdrop for logical inferentialism. This is because natural deduction systems obscure structural elements that are relevant to the meaning of the logical connectives and consecution calculuses make those structural elements explicit.

2.1 Tonk

We begin by highlighting the role of structure in Gentzen-style consecution calculuses. In Gentzen's calculuses, the only structural elements are the comma and the empty sequence, which acts as logical truth on the left and logical falsehood on the right. There are also structural rules, paradigmatically the explicit rules for weakening, contraction, and interchange, identity axioms, and cut.

The example of tonk serves to illustrate the importance of structure and to motivate harmony. These points are brought out, we think, by Belnap's response to Prior's tonk argument.[5] Prior argued that rules alone could not be used to define a connective because there were combinations of rules that trivialized a logic to which they were added, such as the following pair of rules for tonk.

$$\frac{A}{A\text{-tonk-}B} \qquad \frac{A\text{-tonk-}B}{B}$$

Belnap locates the problem differently than Prior does, saying that the introduction of connectives to a system should be seen against the backdrop of an antecedent *context of deducibility*. This context includes the commitment to a certain sort of deducibility relation. The context of deducibility is given by the structural rules of the

---
[5](Belnap, 1962) and (Prior, 1960) respectively.

Philosophical Aspects of Display Logic 287

consecution calculuses. The example of the context of deducibility that Belnap gives is that of Gentzen's structural rules: weakening, permutation, contraction and transitivity. Whether or not a set of rules defines a connective depends on the interaction of those rules with the context of deducibility.

Part of what makes this response illuminating is that it points out that the source of the problem with tonk is its interaction with the structural rules. If one were not committed to a transitive deducibility relation, then tonk need not trivialize the logic. Deducibility in a natural deduction system for classical logic is transitive, and this can be proved as a metatheorem. Natural formulations of such will have the form of the consecution form of transitivity above. Tonk conflicts with something that is obscured in a natural deduction system, which was tonk's original home.

We will now present three examples of structure, obscured in natural deduction, being made explicit in consecution calculuses.

2.2 Example: Discharge functions

Natural deduction systems tend to have little or no explicit structure. A Fitch-style system may have a rule of reiteration. A tree-style system may have no structural rules and little apparent structure. Although not readily apparent, some structure may be found in the arrow introduction rules. The *discharge policies* associated with the arrow introduction rules do a structural job, but they are not found in the rules of the system. The following is an example of a typical $\rightarrow$-In rule.

$$\frac{\begin{array}{c}[A]\\ \vdots\\ B\end{array}}{A \rightarrow B}\;(\rightarrow\text{-In})$$

The brackets indicate that an occurrence of an assumption may be discharged by the rule.

The discharge policies are formal specifications of what assumptions can be discharged when. Classical logic, for example, allows the $\rightarrow$-In rule to discharge zero or more assumptions, as in the following derivation.

$$\frac{\dfrac{\overline{A}}{A \to A}\ (\to\text{-In})}{B \to .A \to A}\ (\to\text{-In})$$

As is common for tree-style formulations, the line over the top formula indicates that it is not an open assumption. The labels on the right indicate which rule is used at each step. The first instance of →-In discharges an assumption while the second does not. In the →−fragment, changing the discharge policies can result in arrows from different logics; banning vacuous discharge, for example, yields a relevant arrow.[6] However, these policies are left implicit in proofs. They do not feature in the rules that are used, and annotated, in many common proof systems.

The structural rules are important for considerations of logical inferentialism, because the meaning, or inferential role, of connectives depends on them.[7] In this example, one arrow rule can be used to produce different sets of theorems when different discharge policies are in place. Because of the importance of structure for the meaning of the logical connectives, the presence of structural rules in consecution calculuses counts heavily in their favor.

2.3 Example: Formula tags

The next example is the use of formula tags in natural deduction for relevance logic. The positive fragment of the relevance logic R can be given a straightforward Fitch-style natural deduction system that uses the device of tags on formulas.[8] The →-In rule removes a tag, which is in the set belonging to the antecedent formula, from the set belonging to the consequent formula. A tag has to be removed, so there cannot be vacuous discharge, in contrast to the previous example. The work done by these tags is captured in the consecution calculus by the rejection of the structural rule K, weakening, and the addition of a new structural connective.[9]

The full logic of R can be given a Fitch-style natural deduction formulation with the appropriate selection of rules for negation. There

---

[6](Hjortland, 2008) goes into this in detail.
[7](Hjortland, 2008) argues decisively for this claim.
[8]See (Anderson & Belnap, 1975) for a canonical formulation of the rules.
[9]Dunn and Mints obtained partial results, formulating consecution calculuses for the positive fragment of R. See (Dunn, 1973) and (Mints, 1976) for details.

is no known Gentzen-style calculus for the full logic of R. The formulation of a consecution system for the full logic of R did not succeed until the adoption of display logic, with its additional structural resources to handle the job done by the natural deduction tags.[10]

The structural job done by these tags is closer to the surface than that of the discharge policies in the tree-style natural deduction systems. The rules for the Fitch-style formulation of R with tags includes the tags in the specification of the rules. The tags are additional structure that is added to the standard Fitch-style introduction and elimination rules. Adjusting how the tags are used, without changing the other parts of the rules, will result in connectives with different inferential behavior. Changing how the tags are used in the rules for the & can alter the inferential behavior of the $\rightarrow$ by upsetting the latter's relevance property. The tags are displayed in the connective rules, but their full contribution is not explicit since there are no rules that govern the tags alone. Display logic provides a consecution calculus for the full logic of R by using mixed sets of structural rules for structural connectives. We will return to mixed sets of structural rules in §3.2.

2.4 Example: Side conditions

The next example is that of modal logic. Modal logics can be given straightforward axiomatic characterizations. Not all modal logics are known to have natural deduction formulations, but the ones that do often have complicated formulations requiring side conditions on modal rules specifying when the rule is applicable. The side conditions do a structural job. An example of a rule with a side condition is a standard formulation of the $\square$-In rule: $A/\square A$, with the side condition that all undischarged assumptions on which A depends are modal formulas. A similar role is played by the restriction of the reiteration rule in Fitch-style systems. In some of those systems, a sentence B may be reiterated within a modal subproof only if B is itself modal.

Gentzen-style consecution calculuses can be given for some major modal logics with no side conditions on rules.[11] Not all modal logics

---

[10] A complete discussion of the role of tags would also need to investigate the Fitch-style formulations of T and E and the various Gentzen-style formulations of the different fragments of T, R, and E, but that is beyond the scope of this paper.

[11] Here and elsewhere we mean that cut-free calculuses can be given, since the inadmissibility of cut is a strike against a consecution calculus.

can be given Gentzen-style consecution formulations because that sort of consecution calculus has few resources for handling the modal elements. Various generalizations of consecution calculuses can handle a wider range of modal logics, without side conditions on rules, by allowing for fine-grained dependency and manipulation of structural elements. In display logic, for example, this can be achieved via an additional modal structural connective governed by separate structural rules.[12]

Structural features are important in determining inferential role. These features are obscured in natural deduction systems in the sense that they do not come out as rules or connectives; rather, they are implicit in the functioning of the natural deduction system. They are made explicit in consecution calculuses. It seems a reasonable principal that a framework which forces one to be explicit about relevant aspects of one's formalism is preferable for philosophical reflection to one that does not. Thus, it is preferable for inferentialism to use consecution calculuses over natural deduction systems. Basing inferentialism on natural deduction systems runs the risks either of fixing on the wrong features of a system or of missing the import of supplementary elements. Natural deduction systems have many virtues, among which is being pleasant to work in. We do not mean here to deny these virtues. However, natural deduction systems obscure the structural details. Consecution calculuses have other virtues, among which is making structure explicit, and for the philosophical purpose of inferentialism this is desirable.

Once discussions of logical inferentialism shift to consecution calculuses, two questions arise. Is one form of consecution calculus preferable? What becomes of harmony? Our answer to the former is that display logic emerges as preferable for philosophical considerations. We will address the latter question by presenting a new explication of harmony suited to display logic.

3 Advantages of display logic

We have argued that consecution calculuses are preferable for inferentialism because they make structure explicit. We will present three advantages that display logic has over other consecution calculuses for

---

[12](Wansing, 1998) develops this form of display logic in detail.

Philosophical Aspects of Display Logic　　　　　　　　　　291

philosophical purposes. First, display logic lets one separate considerations about the logical vocabulary from the context of deducibility in a sharp way (§ 3.1). Second, display logic permits a broader conception of the context of deducibility by mixing structural connectives subject to distinct sets of rules (§ 3.2). Third, display logic is a general framework in which to compare a wide variety of consecution calculuses (§ 3.3).

3.1   Separation of logical connectives and structure

Display logic separates the logical vocabulary from the context of deducibility in a sharp way. It develops this division by using the distinction between structural connectives and formula connectives. The structure is made explicit in the form of structural connectives and rules governing these connectives. One example is Gentzen's empty sequence being represented as the structural connective I, which enters into structural rules.

The separation is clearest when considering the formula connectives. Unlike standard consecution calculuses, display logic has standardized formula connective rules, which have the same form across a range of logics. In these rules, the principal constituents of an inference are displayed as the sole formulas occurring on whichever side of the turnstile they are introduced, as in the following.

$$\frac{X \circ A \vdash B}{X \vdash A \to B} \ (\vdash \to) \qquad \frac{X \vdash A \circ B}{X \vdash A \vee B} \ (\vdash \vee)$$

The formula containing the connective introduced is displayed on one side of the turnstile. In these examples, different arrows and disjunctions are obtained by adopting different structural rules for the $\circ$ connective.

The isolation of principal constituents in display logic effects a cleaner separation of the background structure from the logical connectives than in calculuses in which structural elements can intrude in connective rules, resulting in apparently different rules; for example, in one formulation of intuitionistic logic, the consequents in the connective rules are restricted to single formulas, which makes them appear to be entirely distinct rules from those found in, say, classical logic. If one takes the rules governing a connective to give its meaning, one should want the relation between a connective and structure to be as clear as possible.

3.2 Hybrid contexts of deducibility

As noted, Belnap suggests seeing logical rules against the backdrop of an antecedent context of deducibility, which is specified without appeal to logical vocabulary. In Belnap's article, this is done by Gentzen's structural rules. Updating Belnap's point for display logic, the context of deducibility is given by postulates on structural connectives. This is natural because the ∘ and I of display logic are the analogs of the comma and empty sequence of Gentzen-style consecution calculuses, respectively.

In display logic, structural connectives are indexed according to the structural postulates that govern them. This appears to result in greater flexibility for capturing logics. In relevance logic, for example, one has some structural connectives that obey the classical group of rules and some connectives that obey the relevance group. A similar mixing of rule groups can be seen in Belnap's original formulation of modal logics. Multi-modal logics can also be given a display formulation with distinct • connectives for each modality, which need not all be governed by the same structural rules. This is to say that display logic opens up some relevant possibilities for the inferentialist that are not apparent in a Gentzen-style consecution calculus.

3.3 Generality

In display logic, there is the possibility of having hybrid contexts of deducibility. Other generalizations of the consecution calculus also provide space for more structural combinations and manipulations than are possible in a Gentzen-style calculus. While this might seem to make them attractive to the inferentialist, recent work by Wansing,[13] Poggiolesi,[14] and others[15] indicates that these other generalizations can be adequately simulated in display logic.

Display logic can express the structural combinations that the other generalizations of Gentzen-style calculuses can. Wansing indicates the philosophical import of this, saying, "Display logic might thus serve as a background theory used to compare with each other

---

[13] (Wansing, 1998).

[14] (Poggiolesi, 2008).

[15] Extensive summaries of the work of these other researchers can be found in (Poggiolesi, 2008) and (Wansing, 1998).

various kinds of generalized sequent systems."[16] The preceding two points indicate a virtue of display logic, namely that it allows one to compare many logics within a single, unified framework. The fact that the formula connective rules are the same across a range of logics means that differences in the inferential role of the connectives comes down to difference in structure. These differences can be systematically assessed within the unified framework provided by display logic.

The possibility of having a standard formulation of logical rules that can be used with a hybrid context of deducibility appears to reveal what contribution structural elements can make to the inferential role of logical connectives. It also permits a broader range of logics to be given consecution formulations. Finally it makes structural considerations even more salient, since differences in the meaning of logical connectives are reduced to differences in the context of deducibility. Display logic, like other formalisms examined above, has its virtues and faults. One fault is that it has not yielded decision procedures, so, if that is one's concern, another formalism should be adopted. Having argued that display logic is preferable for logical inferentialism, we will turn the question of harmony.

4 Harmony

The above considerations of display logic suggest a formulation of harmony appropriate for display calculuses:

DL Harmony The left and right rules of a family of connectives are in harmony when it is possible to prove the identity and elimination theorems for them using the proof methods of (Belnap, 1992).

The condition requiring the use of the methods from (Belnap, 1992) means that the proofs provide extra information in that the derivations involved in the proofs of the identity theorem and checking condition eight on analyses do not use structural rules. It is clear what it is to prove the identity theorem and the elimination theorem with the extra restrictions of only using the connective rules and display equivalences. Additionally, it is motivated by general considerations of display logic, since both the identity and elimination theorems arise naturally in that context.

---

[16](Wansing, 1998, p. 187).

It would be an interesting exercise to see which historical examples of unharmonious connectives violate DL harmony and how DL harmony compares with other versions of harmony. We will conclude by looking at what our proposal says about two examples, tonk, which is as unharmonious as can be, and its dual, tunk. Tonk has a left rule like & and a right rule like ∨:

$$\frac{A \circ B \vdash X}{A\text{-tonk-}B \vdash X} \text{ (tonk}\vdash\text{)} \qquad \frac{X \vdash A \circ B}{X \vdash A\text{-tonk-}B} \text{ (}\vdash\text{tonk)}$$

Our proposal, DL harmony, says that tonk is unharmonious. The problem is with the elimination theorem part of DL harmony, more specifically, with the eighth condition on analyses. This can be seen by looking at how the derivations would go. A derivation with full cut be the following.

$$\frac{\dfrac{\dfrac{A \vdash A}{A \vdash A \circ B} \text{ (}\vdash K\text{)}}{A \vdash A\text{-tonk-}B} \text{ (}\vdash\text{tonk)} \quad \dfrac{\dfrac{B \vdash B}{A \circ B \vdash B} \text{ (}K \vdash\text{)}}{A\text{-tonk-}B \vdash B} \text{ (tonk}\vdash\text{)}}{A \vdash B} \text{ (cut)}$$

We cannot obtain the final consecution, $A \vdash B$, using only a restricted form of cut and the display equivalences. Each cut on a subformula A [or B] will introduce additional copies of B [or A] into the consecution. These copies will be preserved by display equivalences, so the conclusion of the original cut will be unobtainable. Condition eight is not verifiable using the methods that provide extra information, so tonk fails to be harmonious.

The dual connective, tunk, has a left rule like ∨ and a right rule like &.

$$\frac{A \vdash X \quad B \vdash Y}{A\text{-tunk-}B \vdash X \circ Y} \qquad \frac{X \vdash A \quad Y \vdash B}{X \circ Y \vdash A\text{-tunk-}B}$$

As one would expect, it is unharmonious because the identity theorem is not provable for it. Inspection of the rules reveals that there is no way to derive

$$A\text{-tunk-}B \vdash A\text{-tunk-}B.$$

Whichever rule is applied first will preclude the application of the other rule.

Shawn Standefer
Philosophy Department
University of Pittsburgh
Pittsburgh, PA 15260 USA
standefer@gmail.com

References

Anderson, A., & Belnap, N. (1975). *Entailment: The logic of relevance and necessity* (Vol. 1). Princeton, NJ: Princeton UP.

Belnap, N. (1962). Tonk, plonk and plink. *Analysis, 22*(6), 130-134.

Belnap, N. (1992). Display logic. In A. Anderson, N. Belnap, & J. M. Dunn (Eds.), *Entailment* (Vol. 2, p. 294-332). Princeton, NY: Princeton UP.

Curry, H. (1977). *Foundations of mathematical logic.* Mineola, NY: Dover.

Dunn, J. M. (1973). A 'Gentzen' system for positive relevant implication (abstract). *Journal of Symbolic Logic, 38*, 356-357.

Hjortland, O. (2008). *Proof-theoretic harmony and structural assumptions.* (Unpublished manuscript)

Mints, G. (1976). Cut-elimination theorem in relevant logics. *Journal of Soviet Mathematics, 6*, 422-428.

Poggiolesi, F. (2008). Display calculi and other modal calculi: A comparison. *Synthese*.

Prior, A. (1960). The runabout inference ticket. *Analysis, 21*(2), 38-39.

Wansing, H. (1998). *Displaying modal logic.* Dordrecht: Kluwer.

Decision Problems of some Intermediate Logics and Their Fragments

Vítězslav Švejdar[*]

1 Tautologies and complexity classes

One of the basic notions in logic is that of *tautology*: a propositional formula A is a (classical) tautology if $\forall v(v(A) = 1)$, i.e. if it has the value 1 (true) under every truth evaluation v. Let, e.g., A be the formula $p \& q \to (\neg r \to \neg p)$, where some parentheses are omitted because we assume that conjunction $\&$ and disjunction $\vee$ have higher priority than implication $\to$ (and equivalence $\equiv$). This formula A is not a tautology because, for the evaluation v such that $v(p) = v(q) = 1$, $v(r) = 0$, we have $v(A) = 0$. The procedure of finding out whether A is a tautology by going through all relevant truth evaluation is known as the *truth table method*.

As an *algorithm*, the truth table method can be analysed from the point of view of its efficiency, i.e. by estimating its time and space requirements. Here time is the time needed to process the given input (given instance of the decision problem), and space is the size of memory needed for auxiliary data when processing the given input. Define *length of a formula* A as the number of occurrences of logical connectives and atoms in A. So parentheses do not count; note however that further computational-complexity considerations show that the exact definition of length does not make much difference. Under our definition of length, a formula A of length n can contain as much as $(n+1)/2$ different atoms, and thus there are $2^{(n+1)/2}$ truth evaluations that the truth table method has to take into account when processing the formula A. Likewise, the difference between the functions $n \mapsto 2^{(n+1)/2}$ and $n \mapsto 2^n$ (and $n \mapsto 10^n$, etc.) does not make much difference; these are functions of exponential growth.

---

[*]This work is a part of the research plan MSM 0021620839 that is financed by the Ministry of Education of the Czech Republic.

Examples of tautologies in elementary logic textbooks usually contain two or three atoms. One of the reasons is that, with four or more atoms, the truth table is unpleasantly long. A formula having 20 different atoms can still fit one single line; the 2^{20} lines of the corresponding truth table is much more than an average (or any) book.

Thus both time and space requirements of the truth table method grow exponentially with the size of the input formula, i.e. the truth table method is an algorithm working in exponential time and in exponential space. While no (essential) improvement in time is known, an improvement in space is possible. This is because rather than writing down all the truth evaluations for the given formula A at once, it is possible to consider only one of them at a time, and reuse the same memory to cycle through all of them. Such improved algorithm works in polynomial space (and exponential time). Since the truth evaluation of a formula A (can be written down so that it) has size not exceeding that of the formula A, it in fact works in linear space. The distinction between linear and polynomial is however not essential for our purpose: both linear and polynomial functions grow slowly in comparison with the exponential function.

The classes of all problems decidable in polynomial time, polynomial space, and exponential time are denoted *P*, *PSPACE*, and *EXPTIME* respectively. The class *P* is often considered to be the class of all efficiently decidable problems. It is clear from the cosiderations above that CPL, the problem to decide whether a given formula is a classical tautology (the letters stand for "classical propositional logic"), is a problem both in *PSPACE* and in *EXPTIME*. In fact, *PSPACE* is a subclass of *EXPTIME*. It is however *not known* whether CPL is in *P*: no algorithm considerably better than the truth table method has been invented, but a proof that no such algorithm is possible also has not been exhibited.

If $A \notin$ CPL, i.e. if a formula A is not a tautology, it might be difficult to find the truth evaluation v such that $v(A) = 0$. However, once the truth evaluation v is given (or guessed), one can verify in polynomial time that $v(A) = 0$. The class of all decision problems with efficient verifiability of positive instances is denoted *NP*. These are problems efficiently decidable by an algorithm that can guess (proceed non-deterministically). Note that such non-deterministic algorithms have no applications in say software development; the notion of non-deterministic algorithm is a theoretical tool for discriminating

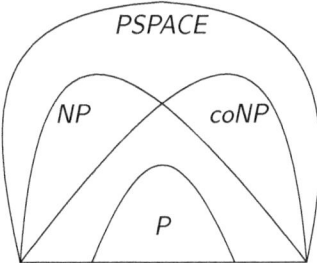

Figure 1: Relations between the classes *P*, *NP*, *coNP* and *PSPACE*

between problems that are not efficiently decidable. As to tautologies, CPL is a member of *coNP*, the problems with efficiently verifiable *negative* instances, since it is not tautologies but non-tautologies that can be recognized using a non-deterministic algorithm. The relationship between the classes *P*, *NP*, *coNP*, and *PSPACE* is shown in Fig. 1. It is believed but not proved that all these classes are different. Even $P \neq PSPACE$ is an open problem.

A problem D in some complexity class is *complete* in that class if every member of that class is, in a well defined sense, reducible to D. Problems complete in a class are the most complex (most difficult) problems in that class. Each of the classes *NP*, *coNP*, and *PSPACE* have complete problems in them; indeed, CPL is complete in *coNP*. Proving completenss of a problem can be seen as establishing fully its algorithmic complexity. Completeness of a problem in one class can be taken as an evidence that the problem is not a member of any smaller class: a *coNP*-complete problem is not in *P*, and a *PSPACE*-complete problem is neither in *NP* nor in *coNP*, in both cases unless two or more classes in Fig. 1 coincide. Such a collapse is not proved to be impossible, but is considered highly unexpected.

Besides CPL, in this paper we will also consider IPL, the decision problem of intuitionistic propositional logic. We will survey results, mostly known, about complexity of IPL and also of some of its subproblems.

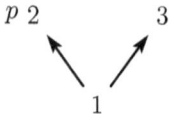

Figure 2: A Kripke model for intuitionistic logic

2 Intuitionistic propositional logic and its complexity

In intuitionistic propositional logic, we deal with the same formulas as in classical propositional logic. They are built from atoms and the symbol $\bot$ for falsity using the symbols $\to$, &, $\vee$, $\neg$.

A *Kripke model* (for intuitionistic logic) is a triple $K = \langle W, \leq, \Vdash \rangle$, where W is a non-empty set, $\leq$ is a transitive and reflexive relation on the set W, and $\Vdash$, the *truth relation* of K, is a relation between elements of W and propositional atoms satisfying the *persistency condition*: if $x \Vdash p$ and $x \leq y$ then $y \Vdash p$. The truth relation uniquely extends to a relation (still denoted $\Vdash$) between elements of W and *all* propositional formulas satisfying the following conditions: $x \not\Vdash \bot$, $x \Vdash A \& B$ iff $x \Vdash A$ and $x \Vdash B$, $x \Vdash A \vee B$ iff $x \Vdash A$ or $x \Vdash B$, $x \Vdash A \to B$ iff there is no $v \geq x$ such that $v \Vdash A$ and $v \not\Vdash B$, and $x \Vdash \neg A$ iff there is no $v \geq x$ such that $v \Vdash A$. We read $x \leq y$ as "y is accessible from x", and we read $x \Vdash A$ as "x satisfies A" or "A is satisfied in x". The elements of W are called *nodes* (sometimes *possible worlds*). One can easily verify that the persistency condition is true for all formulas, not just atoms.

An example Kripke model is in Fig. 2. In this model the atom p is satisfied in the node 2 and not satisfied in nodes 1 and 3, all remaining atoms are nowhere satisfied. In this model we have $2 \not\Vdash \neg p$, since there is a node v accessible from 2, namely 2 itself, such that $v \Vdash p$. Similarly, we have $2 \not\Vdash p \to q$. For still similar reasons, or by the persistency condition, we have $1 \not\Vdash p \to q$. On the other hand, since $3 \not\Vdash p$ and 3 is the only element accessible from itself, we have $3 \Vdash \neg p$. And similarly, $3 \Vdash p \to q$.

A model $\langle W, \leq, \Vdash \rangle$ is a *counter-model* of a formula A if there exists an $x \in W$ such that $x \not\Vdash A$. A formula A is an *intuitionistic tautology* if there is no counter-model of A. Let IPL be the set of all intuitionistic tautologies.

An example of an intuitionistic tautology is any formula of the form $\neg\neg A \to A$. In the model in Fig. 2 we have $1 \not\Vdash \neg\neg p \vee (p \to q)$. Since this formula is a classical tautology, we have IPL $\neq$ CPL. One can easily show that IPL $\subseteq$ CPL and that any formula A is a classical tautology if and only if $\neg\neg A$ is an intuitionistic tautology. The latter fact is known as *Kolmogorov theorem*. This theorem, i.e. the equivalence $A \in \text{CPL} \Leftrightarrow \neg\neg A \in \text{IPL}$, says that CPL is reducible to IPL. Put otherwise, IPL is not algorithmically simpler than CPL. The following theorem says that it is in fact strictly more complicated: *PSPACE*-completeness of IPL implies that IPL $\notin$ coNP (unless $PSPACE = NP = coNP$).

Theorem 1 (Statman, 1979). *IPL is PSPACE-complete.*

For the proof see (Statman, 1979); an alternative later proof in (Švejdar, 2003) might be even simpler. We are not giving the proof in this paper. However, we do give the essential step. It consists in constructing formulas that have Kripke counter-model, but have no small counter-model. Consider the sequence $\{\, D_i \,;\, i \in \mathbb{N}\,\}$ of formulas defined recursivelly as follows:

$$D_0 = \bot, \qquad D_{n+1} = (D_n \to q_n) \to (p_n \to q_n) \vee (\neg p_n \to q_n).$$

The formula D_0 contains no atoms and, of course, is not a classical tautology. Each of the remaining formulas D_{n+1} contains atoms $p_0, \ldots, p_n$ and $q_0, \ldots, q_n$ (only) and is a classical tautology: the subformula $(p_n \to q_n) \vee (\neg p_n \to q_n)$ itself is a classical tautology. None of the formulas D_n is an intuitionistic tautology. This is proved by the following induction. Assume that K_0 is a counter-model of D_n, i.e. a model with an element a_0 such that $a_0 \not\Vdash D_n$. One can assume that a_0 is the least element of K_0 (its root). Then a counter-model of D_{n+1} can be constructed from K_0, its disjoint copy K_1 with root a_1, and one additional element r (new root) as shown in Fig. 3.

In K_0 and K_1, the atoms $p_0, \ldots, p_{n-1}$ and $q_0, \ldots, q_{n-1}$ have their original values, and they are evaluated negatively in r. The atom q_n is everywhere negative. The atom p_n is positive in a_1 (and hence everywhere in K_1) and negative in all remaining nodes, i.e. inside K_0 and in r. From $a_0 \not\Vdash D_n$ and the persistency condition we have $r \not\Vdash D_n$. Thus D_n is nowhere satisfied, and $r \Vdash D_n \to q_n$. There is a node v accessible from r, namely a_1, such that $v \Vdash p_n$ and $v \not\Vdash q_n$; so

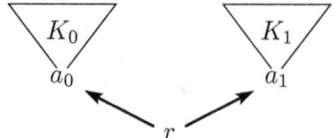

Figure 3: Constructing counter-model of D_{n+1}

$r \not\Vdash p_n \to q_n$. Similarly, there is a node v accessible from r, namely a_0, such that $v \Vdash \neg p_n$ and $v \not\Vdash q_n$; so $r \not\Vdash \neg p_n \to q_n$. Thus indeed, $r \not\Vdash (p_n \to q_n) \vee (\neg p_n \to q_n)$, and $r \not\Vdash D_{n+1}$.

Some more thinking shows that a counter-model of D_{n+1} cannot be much different from the model in Fig. 3: it *must* contain two disjoint copies of a counter-model of D_n. Hence it is at least twice as big. This shows that the formulas D_n are as desired: their sizes grow only polynomially, whereas the sizes of their smallest counter-models grow exponentially.

Statman's theorem and the construction above confirm what one would intuitively think about classical and intuitionistic logic: the latter is algorithmically (strictly) more complicated. This is an instance of a more general phenomenon, observed in various areas of logic. Stronger theories or axiomatic systems, resulting from weaker ones by adding axioms (the reader should feel free to think about, say, set theory and its extensions) cannot be algorithmically more complicated than the weaker ones: the additional axioms forbid something, and their addition simplifies, never complicates, the situation.

There are logics that extend intuitionistic logic but are weaker than classical logic. These logics are called *intermediate*. A natural question reads: where, on the path from intuitionistic logic to classical logic, the PSPACE-complete decision problem turns to the simpler coNP-complete decision problem? Other thing to note is that we need more and more atoms to construct the formula D_n above, and that these formulas contain the connectives $\to$, $\vee$, $\neg$, but no conjunctions. So other natural questions are the following. What happens if the number of atoms is restricted (i.e., fixed)? What happens if the use of logical connectives is restricted?

3 Restricting connectives or the number of atoms

The question whether the decision problem simplifies if the number of atoms is fixed is interesting because in classical logic it does simplify. If, for example, the number of possible atoms is 3, then the number of truth evaluations is 8, and the time needed to check whether a formula A with length n is satisfied by all the 8 evaluations grows only moderately (polynomially) with n. So the decision problem of classical logic with a fixed number of atoms is in P. However, the following theorem shows that intuitionistic logic is different in this respect.

Theorem 2 (Rybakov, 2006). *The decision problem of intuitionistic logic remains PSPACE-complete even if the number of atoms is restricted to* 2.

The case where there is only one propositional atom is also interesting. In classical logic and with one atom p only, there exist only 4 non-equivalent formulas: p and its negatin $\neg p$, their conjunction $p \& \neg p$ (i.e. $\bot$), and their disjunction $p \vee \neg p$ (i.e. $\top$). In intuitionistic logic, the situation is more complicated since, e.g., $\neg\neg p$ is not equivalent to p, and formulas $\neg\neg p \to p$, $p \vee \neg p$, $\neg p \vee \neg\neg p$, $(\neg\neg p \to p) \to p \vee \neg p$ are not intuitionistic tautologies and are not mutually equivalent. There are infinitely many non-equivalent formulas built up from the atom p only. However, these infinitely many formulas form an interesting and rather well-organized structure called *Rieger-Nishimura lattice*, invented independently by Rieger, Nishimura, de Jongh, ..., see Rieger (1949). Closer inspection of properties of this structure shows that this one-atom fragment of intuitionistic logic is in P. As such it can be neither coNP-complete nor PSPACE-complete.

It should be noted that the precise computational complexity status of the one atom fragment of intuitionistic logic might be an interesting problem. LOG is another complexity class, of problems decidable in logarithmic space. LOG is a subclass of P. While it is known that the decision problem of the one atom fragment of intuitionistic logic in in P, it is not known whether it belongs to LOG.

The following theorem speaks about the situation where the use of logical connectives is restricted.

Theorem 3. *The purely implicational fragment of intuitionistic logic, i.e. the set of all intuitionistic tautologies built up from (any*

number of) atoms using implication $\to$ as the only connective, is
PSPACE-complete.

The proof of this theorem is based on the fact that the construction of the formulas D_n, given above, can be improved so that the formulas are built up using implication only. For the full proof see Švejdar (2003). The proof is also implicit (in fact, almost explicit) in the earlier Statman's paper (Statman, 1979).

An interesting question is what happens if both restrictions apply, i.e. if implication is the only connective and simultaneously the number of atoms is restricted. The answer is given by the following theorem, that immediately follows from results in (Urquhart, 1974):

Theorem 4 (Urquhart, 1974). *For any number n, the implicational fragment of intuitionistic logic with n atoms only is decidable in polynomial time.*

In fact, for any fixed n the number of non-equivalent implicational formulas in n atoms, as well as the number of different Kripke models relevant for these formulas, is finite. For example, with one atom p only, p and $p \to p$ are the only two non-equivalent formulas. With two atoms p and q, it is not quite trivial to verify that the number of non-equivalent formulas, like p, q, $p \to (q \to p)$, $p \to (p \to q)$, ... is exactly 14. For more information, see the thesis (Blicha, 2010).

4 Remarks on intermediate logics

A *logic* is sometimes defined as a set of propositional formulas closed under the rules modus ponens and substitution. A logic is *consistent* if it is not the set of all formulas. Consistent logics extending the intuitionistic logic (i.e. containing the set IPL) are called *intermediate*. It can be verified that all formulas in an intermediate logic are classical tautologies. Thus CPL is the strongest intermediate logic. Specific intermediate logic are often defined by adding one or more axiom schemas to intuitionistic logic. For example *Gödel–Dummett* logic G (also denoted LC) is obtained by adding the schema $(A \to B) \vee (B \to A)$. *Testability logic* KC (also called Jankov's logic, or De Morgan logic, or the Logic of weak excluded middle) is obtained by adding the schema $\neg A \vee \neg\neg A$ to intuitionistic logic. Recall that the classical logic is obtained from the intuitionistic logic by adding the schema $A \vee \neg A$ (or equivalently, $\neg\neg A \to A$). A sound argument in intuitionistic logic is this: if $\neg\neg A \to \neg A$ then $\neg A$; if $\neg A \to \neg\neg A$ then $\neg\neg A$. This argument shows that testability logic KC is a sublogic of Gödel-Dummett logic G. Thus from computational complexity point of view, G might be simpler than KC but not vice versa, both logics might be simpler than IPL.

One can easily verify that the logic KC is sound with respect to Kripke models with a greatest element. In fact, a completeness theorem with respect to this class holds. Since $\neg A \vee \neg\neg A$ is not an intuitionistic tautology, KC is stronger than IPL. Since it is easy to construct a counter-model of the formula $(p \to q) \vee (q \to p)$ having a greatest element, KC is weaker than G. The logic G can be verified to be sound with respect to linearly ordered Kripke models, and it is known to be complete with respect to this class.

Since linearly ordered Kripke models correspond to linearly ordered truth values, Gödel-Dummett logic G is studied as one of *fuzzy logics*. It was originally considered in connection with the question whether intuitionistic logic can be characterized as a logic with finite number of truth values, see Gödel (1932). M. Dummett (Dummett, 1959) showed that, in G, disjunction $\vee$ is expressible in terms of the remaining connectives; neither IPL nor KC have this property.

Testability logic KC is discussed in (Gabbay, 1981). This logic is important in connection with the question which completeness theorems (various formulations for various logics) can be proved if the given logic is accepted on metamathematical level (accepted as metalogic) instead of the classical logic. For more on this topic, see Carter (2008).

A linearly ordered Kripke counter-model of a formula A has size not (significantly) exceeding the size of A. So in the logic G, one cannot construct an expansive sequence of formulas similar to the sequence $\{D_n \, ; \, n \in \mathbb{N}\}$ above. This fact has a consequence that G is a decision problem in coNP; thus from algorithmic point of view, G is simpler than IPL. The following theorem says that KC represents the same level of algorithmic complexity as IPL.

Theorem 5. *KC is PSPACE-complete. Its purely implicational fragment is PSPACE-complete as well.*

Proof. To prove this theorem, it is sufficient to show that KC is conservative over IPL with respect to purely implicational formulas. So let A be a purely implicational formula such that $\text{IPL} \nvdash A$. We have to show

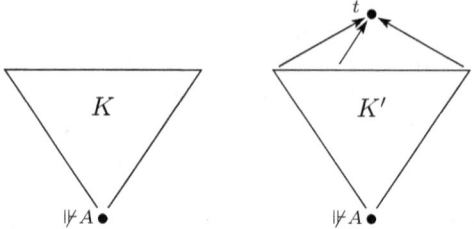

Figure 4: Adding a top node to a Kripke model

that KC $\nvdash A$. Let $K = \langle W, \leq, \Vdash \rangle$ be an intuitionistic counter-model of A. Let $K' = \langle W \cup \{t\}, \leq', \Vdash' \rangle$ be a model as in Fig. 4, constructed from K by adding a new greatest element t, i.e. an alement accessible from everywhere in K. Let all atoms be evaluated positively in t and have the same value in all remaining nodes as they had in K. It is immediate that every purely implicational formula B is satisfied in t. An easy induction shows that for every purely implicational formula B and for each node $a \in W$ we have $a \Vdash B \Leftrightarrow a \Vdash' B$.

The same result is also claimed in (Rybakov, 2006). Out of other popular intermediate logics, see their entry http://en.wikipedia.org/wiki/Intermediate_logic in Wikipedia, Kreisel-Putnam logic and Scott's logic are *PSPACE*-complete as well, because these two logics are sublogics of KC.

It is clear from Theorem 5 and *coNP*-completeness of G that G cannot be conservative over IPL with respect to purely implicational formulas. Indeed,

$$((p \to q) \to r) \to (((q \to p) \to r) \to r)$$

is an example of a purely implicational formula in G which is not an intuitionistic tautology. This example is, of course, derived from the formula $(p \to q) \vee (q \to p)$, and it suggests how it is possible to simulate disjunctions using additional atoms. One can check, using the results in (Urquhart, 1974), that no such example is possible with two atoms only.

Vítězslav Švejdar
Department of Logic, Charles University
Palachovo nám. 2, 116 38 Praha 1, Czech Republic
vitezslav.svejdar@cuni.cz
http://www.cuni.cz/~svejdar/

References

Blicha, M. (2010). *Implikační fragmenty intuicionistické výrokové logiky (Implicational Fragments of Intuitionistic Propositional Logic)*. Bachelor's thesis, in preparation, College of Arts and Philosophy of Charles University, Department of Logic.

Carter, N. C. (2008). Reflexive intermediate first-order logics. *Notre Dame J. Formal Logic*, *49*(1), 75–95.

Dummett, M. (1959). A propositional calculus with denumerable matrix. *J. Symb. Logic*, *25*, 97–106.

Gabbay, D. M. (1981). *Semantical investigations in Heyting's intuitionistic logic* (No. 148). Dordrecht: Reidel.

Gödel, K. (1932). Zum intuitionistischen Aussagenkalkül. *Anzeiger Akademie der Wissenschaften Wien*, *69*, 65–66.

Rieger, L. S. (1949). On lattice theory of Brouwerian propositional logic. *Acta Facultatis Rerum Naturalium Univ. Carolinae*, *189*, 1–40.

Rybakov, M. N. (2006, September). Complexity of intuitionistic and Visser's basic and formal logics in finitely many variables. In G. Governatori, I. Hodkinson, & Y. Venema (Eds.), *Advances in Modal Logic 6 (AiML'06)* (pp. 394–411). Noosa, Australia: King's College Publications, 2006.

Statman, R. (1979). Intuitionistic propositional logic is polynomial-space complete. *Theoretical Comp. Sci.*, *9*, 67–72.

Švejdar, V. (2003). On the polynomial-space completeness of intuitionistic propositional logic. *Archive Math. Logic*, *42*(7), 711–716.

Urquhart, A. (1974). Implicational formulas in intuitionistic logic. *J. Symb. Logic*, *39*(4), 661–664.

www.ingramcontent.com/pod-product-compliance
Lightning Source LLC
Chambersburg PA
CBHW060554230426
43670CB00011B/1822